D0535150

Teh NOV 3 0 2000

609 INV
INVENTING THE MODERN WORLD
: TECHNOLOGY SINCE 1750 /
1ST AMERICAN ED.

DEMCO

inventing the
modern world
technology since 1750

inventing the
modern world

technology since 1750

THE HULTON GETTY PICTURE COLLECTION

SCIENCE & SOCIETY PICTURE LIBRARY

Robert Bud, Simon Niziol,
Timothy Boon and Andrew Nahum
of the Science Museum

A Dorling Kindersley Book

Authors' acknowledgments

To cover the history of modern technology in a short text has required as much help and advice as a much longer work. The authors are grateful to all those who so willingly read and criticized drafts and extracts, challenging facts and balance as well as linguistic style, and then offering solutions.

Early drafts were read by Lindy Biggs, Ian Inkster, Bill Luckin and Seymour Mauskopf. We are also grateful for the advice of Brenda Lees of the North Highland Archive, and the help of Nino Makhviladze. Staff at the museum were also most generous with their time. The authors must thank Brian Bowers (even after nominal retirement), Roger Bridgman, Peter Fitzgerald, Jane Insley, John Liffen, Robert C. McWilliam, Alan Morton, Susan Mossman, John Ward, David Woodcock and Michael Wright, who were perhaps pestered the most. Other staff at the museum – David Rooney, Claire Seymour and Xerxes Mazda – contributed draft texts at various stages. The staff of the Science Museum Library have contributed research and sourced the most obscure facts. John Griffiths and Alice Nicholls provided invaluable management support of a complex process and Jane Davies has made sure that the team has managed to communicate over many sites.

A book such as this relies on its illustrations. The first debt we owe is to Ali Khoja at the Hulton Getty Picture Collection and to Christopher Rowlin at the Science & Society Picture Library who explored the depths of the holdings with expertise and imagination. Within the Science Museum new photography was carried out by David Exton, Claire Richardson and Ron White. The book relies too on a wide range of sources beyond the two major holdings. Primarily we are grateful, again, to Ali Khoja, for his persistence and discrimination and to Joelle Ferly, to Jenny Speller and to Suzanne Hodgart for further picture research.

Editing is at the heart of the process. We have been fortunate for the support and detailed assistance of Doron Swade, Assistant Director and Head of Collections at the Science Museum and for the goodwill and scholarship of Roger Hudson. Paul Welti has been an endlessly tolerant designer willing to incorporate the most inexcusable changes. Mary Osborne has been almost equally flexible as a project manager, though adamant when necessary. At the Science Museum, Angela Murphy, Head of the Science & Society Picture Library, and Ela Ginalska, Head of Publications, were invaluable facilitators. Finally, this book owes its existence to the instant willingness of Charles Merullo, at Hulton Getty, to take on an idea not just for a book but for a collaboration between two complex institutions.

Dorling **DK** Kindersley

LONDON, NEW YORK, SYDNEY, DELHI, PARIS, MUNICH, and JOHANNESBURG

Project Editor Barbara Berger
Additional Editorial Crystal Coble
Cover Design Dirk Kaufman
Picture Research Ali Khoja, Christopher Rowlin and Joelle Ferly
Editor Roger Hudson
Design Paul Welti
Production Mary Osborne
Typesetting Peter Howard

First American Edition, 2000
2 4 6 8 10 9 7 5 3 1
First published in the United States by
DK Publishing, Inc.
95 Madison Avenue
New York, NY 10016

Text copyright © Board of Trustees of the Science Museum, 2000

see our complete catalog at
www.dk.com

This book was created, designed and produced by the Hulton Getty Picture Collection, 21–31 Woodfield Road, London W9 2BA Fax 44 (020) 7266 2658 in conjunction with the Science Museum

All rights reserved by the Hulton Getty Picture Collection and the Science Museum. Without limiting the rights under copyright reserved above, no part of this publication may be reproduced, stored in or introduced into a retrieval system, or transmitted, in any form or by any means (electronic, mechanical, photocopying, recording or otherwise), without the prior written permission of both the copyright holder and the publisher of this book

Library of Congress Cataloging-in-Publication Data

Inventing the modern world : technology since 1750 / by the staff of the Science Museum.– 1st American ed.
    p. cm
  Includes index
  ISBN 0-7894-6828-X (acid-free paper)
   1. Technology–History

  T18 .I565 2000
  609–dc21

                  00-026045

Origination by Omniascanners srl, Milan, Italy
Printed by Nuovo Istituto Italiano d'Arti Grafiche SpA, Bergamo, Italy

(Above) **When television was new** Demonstrating home television reception, November 1936

(Half title page) **Charles Babbage's Analytical Engine** Trial assembly made for the never-completed predecessor of the modern computer, c. 1865

(Title page) **Modern computing** Electronic circuitry for a computer hard disk drive, c. 1998
Crash test dummy Adult female, model type 'Sierra Susie', used by the Motor Vehicle Industry Research Association, c. 1970

Contents

Sci√m
science
museum
Science & Society Picture Library represents the collections of the National Museum of Science & Industry (NMSI) which comprises the Science Museum in London, the National Railway Museum in York, and the National Museum of Photography, Film & Television in Bradford. SSPL documents the history of science, technology, transport and the media.

HG HULTON GETTY
The Hulton Getty Picture Collection is one of the greatest resources of retrospective photography in the world. There are over 300 different collections, ranging from photojournalism to prints and engravings. Hulton Getty is part of Getty Images, the largest supplier of visual content in the world.

Preface

Daily, technology plays a greater part in our lives. Consumers devour its products and mature economies thrive on it. But there are costs, such as the impact on the environment, and, at the start of the third millennium, we see a growing love-hate relationship with high technology. The study of how technology has fared at other times in our history continues to be fascinating – and profoundly revealing.

Inventing the Modern World portrays 250 years of change in Western technology beginning with the Industrial Revolution of the 18th century. Its focus is mainly on Great Britain, the world's first industrial nation, but it also encompasses the seminal American and German contributions to the Second Industrial Revolution of the late 19th century, the impact of two world wars, and the late 20th century phenomenon known as globalization. *Inventing the Modern World* recounts technical achievements, certainly, but also the way these were received, and features the views of those who expressed doubt as well as those who promoted new technologies.

Inventing the Modern World grows out of an original analysis of technological change underpinning an ambitious new gallery, "Making the Modern World," at London's Science Museum. "Making the Modern World" is the largest single historical panorama in the Museum's history. However, the volume stands on its own as a conjunction of current scholarship and powerful, intriguing and arresting images. Some of these, such as a photograph of spectators at a 1911 aviation meeting on pp. 124-5, are relatively unknown. Others, like the paintings of J. M. W. Turner, are more familiar, but not, perhaps, in the context of the history of technology. These works express a 19th century fascination with new technologies tinged with sadness at the loss of the old.

The almost 500 images included in *Inventing the Modern World* come, mainly from two rich resources: the Science Museum's Science & Society Picture Library, built up over almost 150 years, and the Hulton Getty Picture Collection, which ranges from 19th century exhibition records to the vast photo archive of *Picture Post* magazine.

Inventing the Modern World is offered as a tribute to the astonishing accomplishments it portrays and as a reflection on the relationship between technical change and industry, between science and technology, and between people and artifacts.

Coalbrookdale by Night
An archetypal image of the Industrial Revolution: the oil painting by Philippe Jacques de Loutherbourg (1801) shows the buildings of an iron works against the glow from the furnace being tapped.

While the French Revolution carried out its great experiments on a volcano, England began its own on the field of industry

ADOLPHE BLANQUI
History of Political Economy in Europe, 1837

Inventing accuracy

Until well into the 20th century, the story of technology was divided, as if by a meridian, by the "Industrial Revolution." The term, itself evoking images of violent discontinuity, described changes in the late 18th and early 19th centuries, seen as causing a rapid and thorough transformation of economy and society. Before, people lived on farms, in villages and market towns; now, there were factories and industrial towns. Before, brawn, human and animal, had been the mover; now, it was the steam engine. Before, craft had determined the way things were made; now, it was science. Indeed, one of the first uses of the term by the French historian and economist Adolphe Blanqui in his *History of Political Economy in Europe*, published in 1837, was intended to draw a parallel with the momentous French Revolution of 1789.

John Troughton's dividing engine. Originally the scales of every scientific instrument had to be marked out or "divided" by highly skilled craftsmen. From the 1780s the dividing engine could be used to produce the scales of instruments of moderate size mechanically. This was faster and more consistently accurate (right and above).

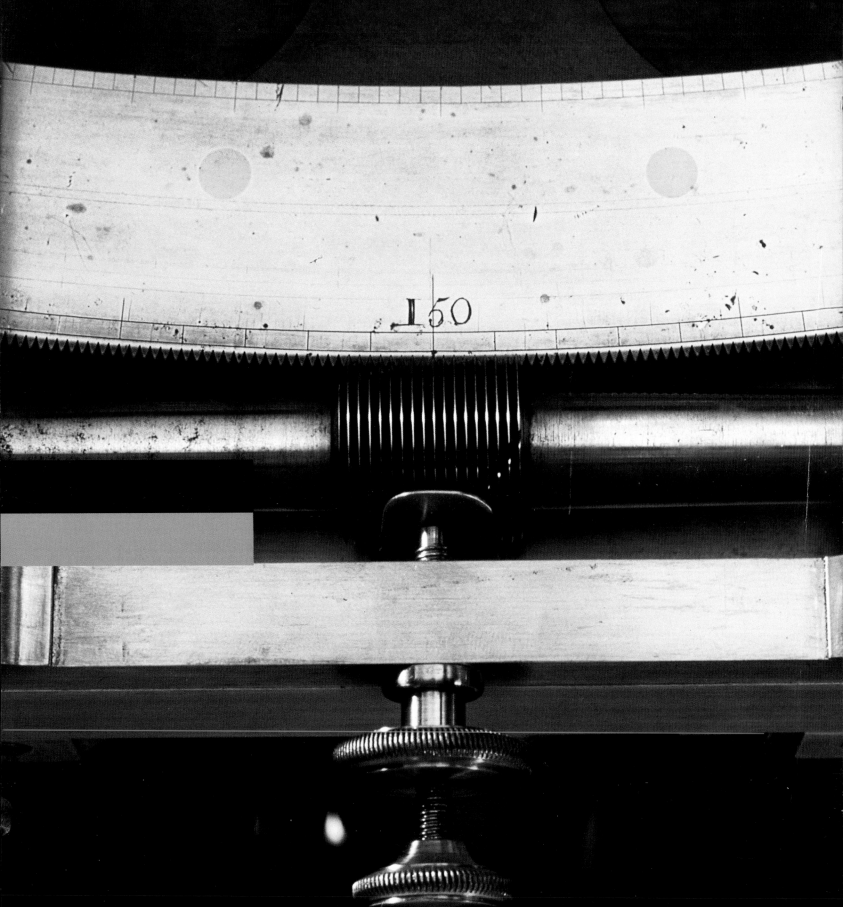

Ironbridge

Crossing the River Severn at Coalbrookdale in Shropshire, England, is the first cast-iron bridge in the world, erected in 1779. The components had been cast at the nearby Coalbrookdale foundry where the smelting of iron with coke from coal rather than charcoal from wood had been pioneered in 1709. Visitors flocked to see the bridge. It also became a favorite subject for artists. This colored aquatint was published in 1823.

Some writers have down-played the suddenness of change and the universality of its effects at the time. Few deny, however, that the developments of the period had a profound effect on subsequent history. It is perhaps, therefore, best to think of the Industrial Revolution not so much as a single event but as a decisive stage of an ongoing process which continues to this day. The cumulative force of changes in the way things were made and the way people worked spread within Britain and then across

Europe and North America. Technological change became the accepted norm, leading to sustained economic growth and social change.

The main technological developments associated with the Industrial Revolution included cheaper ways of producing iron, the use of steam engines and water power for factories and a growing ability to engineer new machines for production. In the places where the new factories were established, the changes were dramatic. In 1776 the pioneering agricultural journalist Arthur Young described the ironworks of Coalbrookdale as "horribly sublime," thus putting them on a par

Engineering genius, scientific instrument and practical achievement

James Brindley with a theodolite, and the Barton aqueduct behind, carrying Brindley's Bridgewater Canal across the river Irwell (above). The arresting spectacle of water over water and boat over boat was one of the wonders of the day (right). The canal, completed in 1776, went from the Duke of Bridgewater's collieries to Manchester, where coal prices were halved. By 1800, a generation before the railway, 2,000 miles of navigable waterways linked Britain's major industrial districts.

with the newly fashionable Peak and Lake Districts (see pages 6 and 7). "Cottonopolis," Manchester and its surroundings in northwest England, was a "must-see" for foreign visitors early in the 19th century. To Benjamin Disraeli, the future prime minister, the city was the wonder of the modern world, "as great a human exploit as Athens." Nevertheless, in most parts of Britain, development was far from sudden. Even in 1851, when Britain was already widely regarded as the "workshop of the world," agriculture still employed far more people than any one of the fast-growing manufacturing industries.

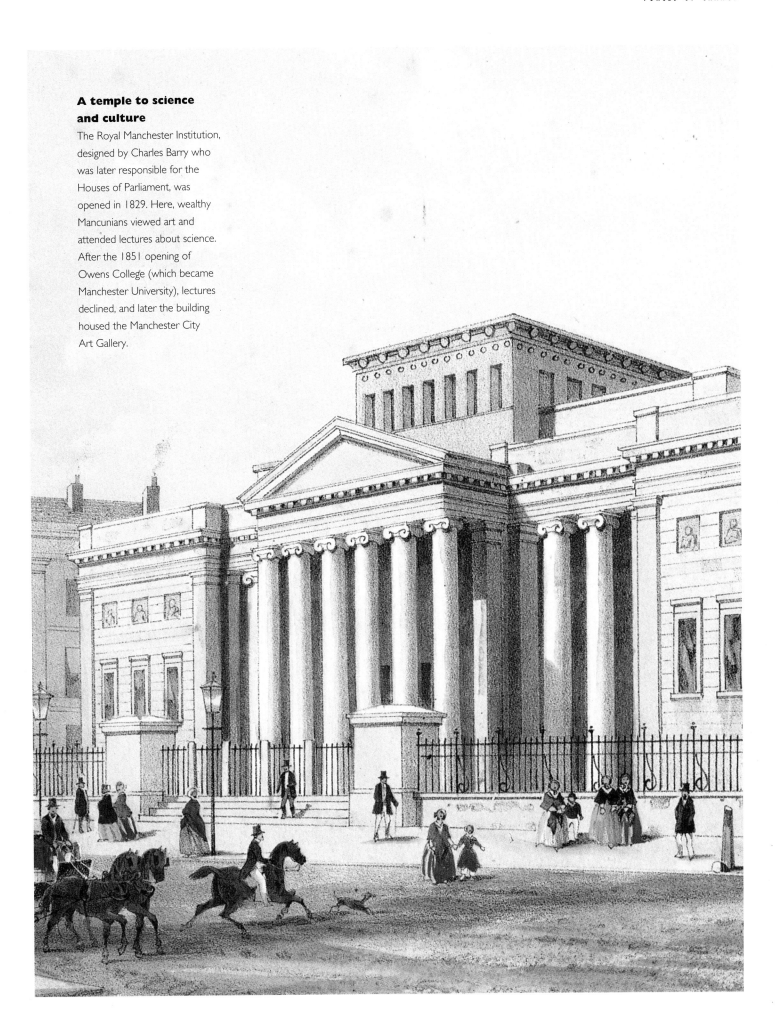

A temple to science and culture

The Royal Manchester Institution, designed by Charles Barry who was later responsible for the Houses of Parliament, was opened in 1829. Here, wealthy Mancunians viewed art and attended lectures about science. After the 1851 opening of Owens College (which became Manchester University), lectures declined, and later the building housed the Manchester City Art Gallery.

The Davy safety lamp

Increasing demand for coal meant the digging of ever deeper mine shafts and a greater risk of explosions caused by the methane gas (fire-damp) being ignited by the naked flames of traditional lamps. The gauze around the flame prevented the gas from exploding and a change in the length and color of the flame also provided early warning of fire-damp. The lamp came to symbolize not only the application of science to the problems of technology, but also technology in a humanitarian rather than a purely economic role.

The contribution of science

Technology and science are often run together. Sometimes it may seem obvious that one leads to the other: atomic physics may seem to have led inexorably to the atom bomb. But as one peers deeper into the details, the relationship between pure scientific knowledge and technical development is often far from clear. During the Industrial Revolution, those with scientific expertise in what was called "natural philosophy" or "philosophical chemistry" did make certain practical contributions. For example, Humphry Davy, a chemist, invented the "safety lamp," which prevented catastrophic explosions of methane in deep coal mines. Developments in the early chemical industry owed much to science. On the other hand, few of the men associated with the breakthroughs in engineering or metallurgy had any formal scientific training, and technological progress often occurred before the relevant scientific principles were explained.

One practical contribution of science does stand out above all others – the search for accuracy. Practitioners of astronomy, navigation, mining and surveying, seeking as they were to define our relationship with space and time with greater precision than ever before, required instruments of the highest quality. Astronomical observations and the elucidation of the size and shape of the earth, as well as the aiming of guns and the driving of mineshafts, relied on making accurate angular measurements. This explains the contemporary interest in the accurate division of the circle for mathematical instruments and the efforts expended in the development of the "dividing engine" to speed up the manufacture of such instruments.

The increasing need for accurate maps demanded surveying instruments as well as standards of length of the greatest precision. The newly improved theodolite, combined with the technique of triangulation, allowed the production of superb county maps containing a wealth of information. Such developments made possible the planning and construction of new canals and roads even in a wilderness such as the Scottish Highlands. In 1784 General Roy began his survey, which was to fix the positions of the observatories at Greenwich and Paris, starting with an accurately measured line just five miles in length outside London. From it a web of triangles was extended across the country, then linked with France before spreading around the world.

Latitude and longitude

The octant (left), invented in 1731, allowed the measurement of angles between distant objects. This allowed the determination of a ship's latitude from the relative position of sun or pole star. The artist has taken liberties here, enlarging the instrument's size. An accurate calculation of longitude was not possible until John Harrison (above) perfected his marine chronometer. He received a large cash prize for his achievement, twenty-seven years after he had developed his first chronometer (right).

The major European powers of the 18th century, with colonial possessions scattered around the globe and trade routes in every direction, were confronted also by the need to chart coastlines and to navigate at sea. The octant, and its successor the sextant, allowed latitude to be determined from a measurement of the sun's height at noon. Longitude was more difficult and many governments offered large prizes to encourage the search. One option for determining it involved a complex astronomical system which required taking a number of measurements of the position of the moon relative to the stars combined with lengthy calculations using specially prepared tables (the "lunar distance" method). Such measurements were difficult to take accurately from the heaving deck of a ship at sea. Another solution required an accurate timepiece. At the outset, making a clock which could cope with the motion of a ship and the variations of climate met on a voyage seemed, to many, an impossible task. The problem was eventually solved by John Harrison whose chronometer of 1759 could record time on board ship, in heat and cold, to within three seconds per day.

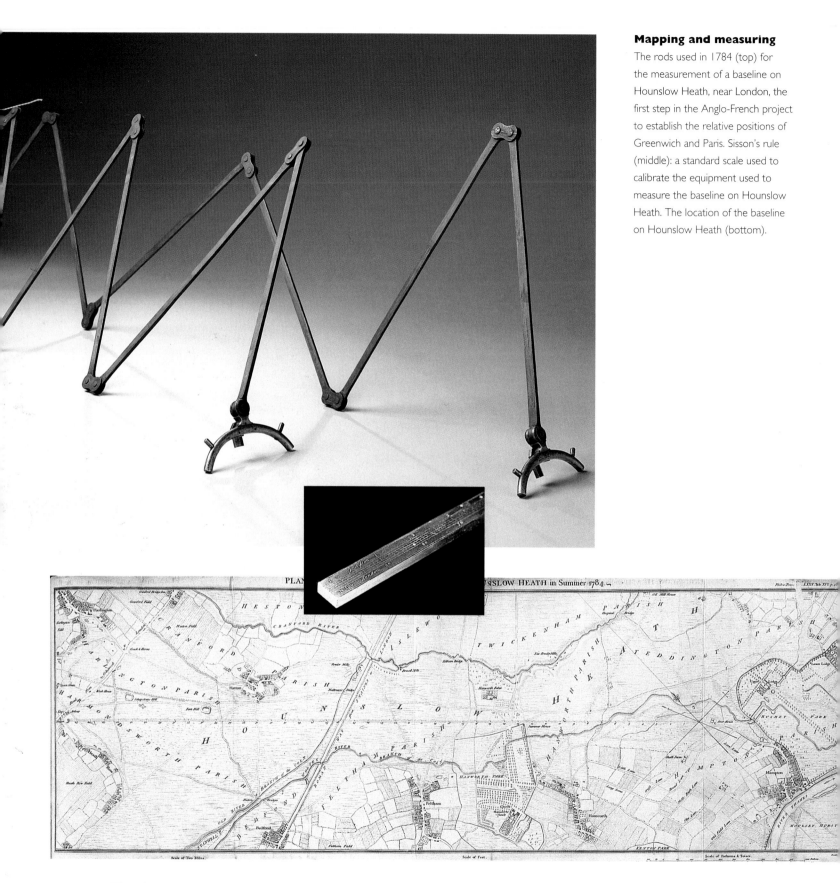

Mapping and measuring
The rods used in 1784 (top) for
the measurement of a baseline on
Hounslow Heath, near London, the
first step in the Anglo-French project
to establish the relative positions of
Greenwich and Paris. Sisson's rule
(middle): a standard scale used to
calibrate the equipment used to
measure the baseline on Hounslow
Heath. The location of the baseline
on Hounslow Heath (bottom).

The drive to measure and define was applied on both a large and a minute
scale: after the French Revolution, the meter – defined as 1/40,000,000 of the
diameter of the earth – was established as a standard unit of length, while,
responding to the needs of engineers such as Henry Maudslay (p.41), micro-
meters were being devised which could read to a thousandth of an inch.

Ramsden's theodolite

The altazimuth theodolite was a
surveying instrument, able to
measure both bearings and
altitudes from a single reading,
perfected during the 18th
century. A theodolite of the
highest quality was required to
measure the angles for the
Primary Triangulation, the project
to map the whole of Britain.
Jesse Ramsden, the inventor of
the dividing engine, duly obliged.
His first theodolite of 1787
offered unprecedented accuracy
– it was even able to measure
the curvature of the earth. The
1791 instrument pictured right
was even more accurate, and
remained in constant use for
over half a century.

Britain enjoyed several advantages over many of its European rivals in terms of internal security, trading networks, overseas possessions, an absence of internal tariffs, legal protection of property, availability of credit, and agricultural productivity. But among its most precious assets was an ample supply of skilled engineers and mechanics. The combination of these favorable factors and an obsession with improvement ensured that Britain reaped the fullest rewards of technological change, even in the case of inventions originating elsewhere. The Jacquard loom, the flax spinning machines of Philippe de Girard, the Leblanc

The Jacquard loom
was introduced for use in the French silk industry in 1801 (top left). It began to appear in numbers in Britain in the 1820s. It allowed the automatic weaving of patterns into fabric, with the information recorded on punch cards using binary code. It was later modified to weave worsted and high-quality cottons. The concept of punch cards was to find many uses elsewhere.

Cotton

was one of the most visually striking of the new industries. Huge mills appeared, often in rural locations (left). Inside, the scale of the new spinning machinery was equally awesome (above). Both illustrations are of the Swainson and Birley Mill in Preston, Lancashire, in 1834.

soda process, Berthollet's chlorine-bleaching method, Koechlin's dyes – these were among the many ideas conceived on the Continent but developed and perfected in Britain. This pattern was to change as the 19th century went on.

The pursuit of improvement paid great dividends, as ever more precise machine tools allowed the construction of new machines of ever greater complexity. Unlike Leonardo da Vinci, whose visionary plans were doomed to remain on paper, practical inventors during the Industrial Revolution could reasonably expect to see their drawings turned into working machines. Furthermore, a culture of observation and experiment produced a steady stream of minor improvements which, though in themselves often trivial, made for decisive gains in efficiency and performance.

Scientific societies

In Britain, remarkable groups of men, who at work were making contributions to industrial development, also gathered in their spare time to discuss science and what were known as "the useful arts." These clusters of enthusiasts formed creative hubs and, in a sense, were the university science and engineering departments of the day. The most famous was a group in the English Midlands known as the Lunar Society, which met at the full moon. In it, James Watt, pioneer of the steam engine, and Josiah Wedgwood, whose factory-made pottery became world-famous, rubbed shoulders with the chemist Joseph Priestley who discovered oxygen, and Erasmus Darwin, grandfather of the even more famous Charles and himself interested in evolution.

In Manchester, the Literary and Philosophical Society founded in 1781 counted many distinguished early members. The best remembered is John Dalton, who was the first to conceive of atoms as balls, each with a weight characteristic of its element. Dalton's young protégé was the brewer James Joule, who is remembered today by the unit of energy – a concept to which he made a major contribution. Such societies were to be found in major provincial centers such as Edinburgh and Newcastle as well as in smaller towns such as Derby, Doncaster and Warrington. Britain was not alone. The American Philosophical Society in Philadelphia, which had roots in a body founded in 1743, included Benjamin Franklin among its members. The American Academy of Arts and Sciences, in Cambridge, Massachusetts, was founded during the American Revolution in 1779.

Britain did not experience anything akin to the French Revolution, but riots were common and there were moments of high tension between town and country, and between masters and men, on several occasions in the early 19th century. Science, with its commitment to objectively observing and exploiting the natural world, seemed a means of bringing together men and women of different faiths and politics.

Josiah Wedgwood

ran his pottery like a factory, with strict division of labor and a steam engine from James Watt. He was a promoter of the Grand Trunk Canal and experimented with materials, glazes and neo-classical designs.

American Philosophical Society

Its seal depicts a Native American being introduced to Minerva, the goddess of learning (left). Benjamin Franklin, American scientist, inventor, statesman, man of letters and a founder of the Society (below).

By the 1830s, Manchester boasted three venues for scientific lectures almost next to each other: a Royal Institution for the wealthy, an Athenaeum for the younger set and a Mechanics Institute for clerks and artisans. Other clubs in other towns, such as Edinburgh, were remarkably inclusive by comparison. They attracted scholars, peers and land owners with an interest in "improvement," as well as manufacturers and even skilled artisans.

In London there were several distinguished organizations. The Society of Arts, founded in 1754, promoted innovation by offering prizes. It continues to flourish, as does the Royal Institution, founded in 1799. There, businessmen, estate owners and high society were lectured on scientific topics, some pure and others applied, by professors such as Humphry Davy, who developed agricultural chemistry, and his successor, the great chemist and physicist Michael Faraday. Other groups included the Mineralogical Society, of which the chemical manufacturer Luke Howard was a member. Though

Michael Faraday

an exemplary promoter of science, lectured at the Royal Institution for over 30 years. He advocated that a lecturer should appear "easy and collected" and stressed the importance of good delivery. He offered the public an attractive mixture of education and entertainment, and inaugurated several series of popular lectures which thrive to this day.

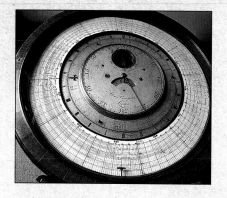

famous primarily for his classification of clouds, Howard was also the proprietor of the first company in Britain to make quinine, the new anti-malaria drug first extracted from cinchona bark in Paris.

The interests of these groups touched a great deal on practical applications but they were not in themselves devoted to what we would easily recognize as "applied science." Instead they developed a culture of learning, of study and observation, of curiosity and interest in the natural world, which seemed appropriate to the newly emerging society.

Luke Howard's barograph clock

(left), which measured and recorded changes in air pressure, and (right) part of a circular graph on which its daily readings were plotted for a whole year. Howard is regarded as the founder of modern meteorology, and had a wide range of scientific interests. He kept a detailed record of the London weather over a 25-year period and proposed theories to explain long-term climatic changes.

Cloud classification

in many important respects is owed to Luke Howard's careful observation of clouds over many years. His terminology and symbols are still largely used today. His watercolors combine precise renderings of clouds (below) with stylized landscapes and stock figures (left).

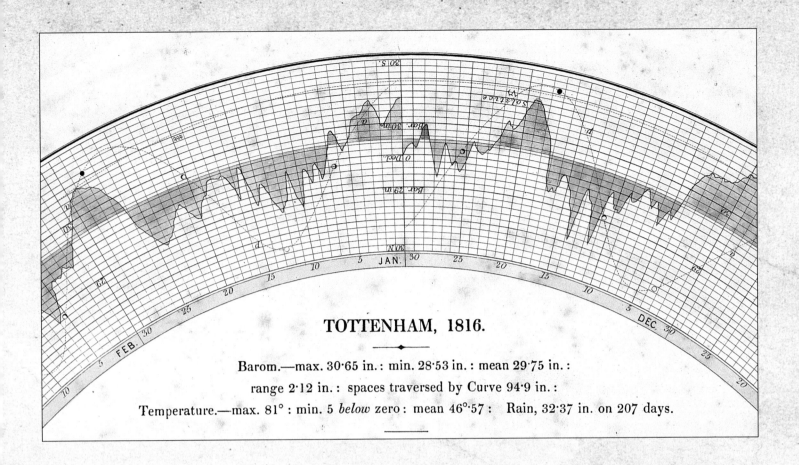

TOTTENHAM, 1816.

Barom.—max. 30·65 in. : min. 28·53 in. : mean 29·75 in. :
range 2·12 in. : spaces traversed by Curve 94·9 in. :
Temperature.—max. 81° : min. 5 *below* zero : mean 46°·57 :   Rain, 32·37 in. on 207 days.

Ficus superba

A 1782 watercolor of a species of fig. This was based on one of the huge number of illustrations of plants made for Joseph Banks during his epic voyage to the South Pacific with Captain James Cook (1768–71).

Sir Joseph Banks

(top) was largely responsible for the enhanced reputation of the Royal Botanical Gardens at Kew (above). He was President of the Royal Society for over 40 years, and served on several government committees concerned with science, trade and agriculture. The Palm House at Kew, seen here, is an example of a technology-led building, its appearance determined by the new materials – plate glass and iron – used in its construction (1848). This type of Victorian high tech building reached its peak with the Crystal Palace (1851).

Sir Joseph Banks, who travelled with James Cook to the South Seas at the end of the 1760s, became the most prestigious English man of science in his day. Not just an observer, he arranged that breadfruit be brought to the West Indies from Tahiti aboard the HMS *Bounty*. He also established the fame of Kew Gardens, near London, as a center of practical botany as well as beauty. Colonies provided opportunities for systematic study of the natural world and of unfamiliar people, with little distinction made between scholarship and exploitation. It was to feed slaves on plantations that breadfruit were brought to the West Indies. The contemporary idea of "improvement" encompassed a belief that almost all problems in the fields of manufacture, agriculture, and even criminal correction, could be addressed by the application of science. Rational entertainment, with both a strong moral and a social dimension, could be obtained along the way.

Alternative approaches

Developments in Britain aroused keen interest elsewhere, and the new manufacturing districts attracted a wide range of visitors from abroad. Some came simply out of curiosity, but others were determined to acquire information which could be of use back home. Many manufacturers and politicians in Britain saw such attempts as a challenge to the country's newly acquired position of strength, and laws were passed to prevent the spread of new technologies by restricting the movement of skilled workers and machines. Foreign visitors were thus frequently obliged to resort to bizarre methods – workers were bribed or plied with drink to divulge details of production processes or allow access to factories.

Technological ingenuity was not a British monopoly, and by the time the last of the legal prohibitions was repealed in 1843, many countries were well on the way to establishing competitive manufacturing industries of their own. This was a protracted process as the results achieved in Britain could not easily be replicated where traditions, markets, natural resources and levels of prosperity were very different. Rapid progress occurred where skills were abundant, as in Belgium and Switzerland. The United States shared

Faraday's laboratory
at the Royal Institution. Here he conducted groundbreaking research in chemistry and physics. He made crucial contributions to the study of electricity and magnetism, discovering the principle behind the electric motor, electric generator and the transformer. Most of his research was conducted alone or with the aid of a single assistant.

Liebig's laboratory

at the University of Giessen in Germany, Justus Liebig, professor of chemistry from 1828, conducted research which established the relationship between chemistry and soil fertility and pioneered the research training of student chemists. The picture depicts the many researchers from all over Europe and North America studying at the laboratory in 1840.

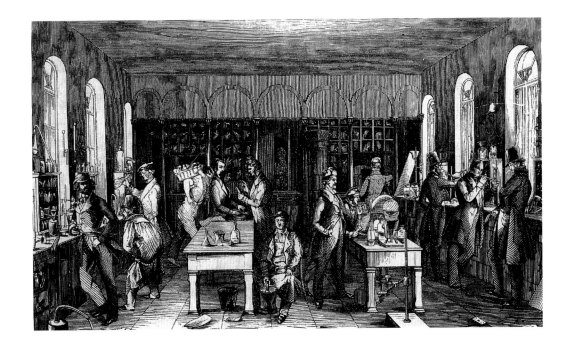

with Britain an obsession with progress and pragmatic improvement, and soon made its own effective contributions to manufacturing.

Most European governments preferred an approach that was less *ad hoc* than Britain's, and more systematic training often emphasised theoretical knowledge. In France and the German states, formal institutions for the creation and dissemination of knowledge were seen as the best way forward. At places like the Ecole Polytechnique in Paris or Saxony's Freiberg Bergakademie, future military or mining engineers were taught by men of science. The contrasting approach encouraged a search for different solutions: whereas French engineers tended to build bridges on scientific principles, their British counterparts used more empirical methods.

In time, as major breakthroughs resulted from a more direct application of scientific knowledge, this ushered in what became known as the "Second Industrial Revolution," characterized by the new chemical and electrical industries.

Seguin's locomotive

of 1829, the first to be built in France, was notable for its use of the multitubular boiler, which Marc Seguin had conceived independently of Robert Stephenson, who also introduced it into the workings of the *Rocket*. This device radically increased the heating surface in contact with the water in the boiler.

Industrial work, still under bondage to Mammon, the rational soul of it not yet awakened, is a tragic spectacle. Men in the rapidest motion and self-motion; restless, with convulsive energy, as if driven by Galvanism, as if possessed by the Devil...

THOMAS CARLYLE
Past and Present, 1841

2 Manufacture by machine

What made the Industrial Revolution seem special to its contemporaries was the growing use of machines in industrial operations and their grouping within factories, to provide unprecedented productive power. The waterwheel and the lathe were known in the ancient world, and manual methods of manufacturing were to survive in many industries into the 20th century. However, the Industrial Revolution transformed the applications of machinery, with radical consequences for the way things were made. Machines could not only replicate the actions involved in manual labor but also carry out tasks far beyond the strength or endurance of human workers.

Arkwright's water frame of 1769 (right and above). Cotton threads were drawn out by pairs of rollers, and then twisted, producing strong coarse yarn. This complemented the finer yarns produced by Hargreaves' spinning jenny and by traditional spinning wheels, and made possible the weaving of all-cotton cloths.

The mechanization of the textile industry

The cotton industry was the first to change in the dramatic way that would become typical of the new order. Cotton was more suitable for manipulation by machine than older-established materials such as wool and silk. The British cotton industry grew at a rapid rate, with the consumption of raw cotton increasing roughly 150 times between the 1770s and the middle of the 19th century. A large number of machines, including many of the classic inventions of the Industrial Revolution – Hargreaves' spinning jenny, Arkwright's water frame, and Crompton's mule – helped transform what had once been a minor branch of textile production into the most valuable single industry in Britain. So cotton, indeed, became king.

Crompton's mule

of 1779. Samuel Crompton spent a decade developing a hybrid machine (hence the name) combining elements of the water frame and the spinning jenny. His mule produced a strong, fine yarn and, with continuous modifications, became the backbone of mechanized spinning for over a century. A contemporary illustration (left), and a later replica (above).

Mule spinning

at the Swainson and Birley Mill, 1835 (above). Speeds of rollers, spindles and carriage could all be adjusted to vary the type of yarn. The biggest rise in productivity came in such factories driven by water or steam. The machine on the right is stationary while a girl appears to be mending a broken thread, and a boy sweeps up under it.

Mechanized weaving

proved more difficult to achieve than spinning. Although powerlooms had been invented as early as the 1780s, it took another 40 years before effective machines became available. Like the mule, the powerloom was constantly modified. The Harrison and Son loom from the 1850s incorporated an emergency stop mechanism (right).

Because of the nature of the processes involved, some stages of production were mechanized with more difficulty than others. So, for example, the making-up of finished garments had to wait for the appearance of the sewing machine, nearly a century later. On the other hand, the spinning of thread into yarn and the weaving of yarn into cloth were

revolutionized sooner. The repetitive action of moving parts could eliminate human error or make up for lack of skill by insuring an unprecedented regularity of quality. Single power sources, the waterwheel and increasingly the steam engine, made it economical to operate larger machines and ever more specialized equipment. The increase in production in textile mills was dramatic.

Coventry ribbons

Change in the silk industry was more modest than in cotton and woollens, and traditional centers of production like Coventry in the Midlands, retained their importance. A Coventry ribbon loom (left); ribbons made in Coventry (above).

International emulation

The large-scale production of cotton goods was one of the first features of the British Industrial Revolution to develop elsewhere. In the United States, Lowell in Massachusetts emerged as the chief manufacturing center. A label used by the Merrimack Manufacturing Company (below).

Powerloom weaving

at the Swainson and Birley Mill in 1835. Handlooms were still widely used, and such scenes would still have been far from typical at this time (left). The man appears to have a broken drive belt in his hands.

The cotton which would have taken two people their entire lives to spin using traditional techniques could be turned into thread in just a few months by a single operator of the new device called a "mule." In turn the development of the virtually automatic "self actor" in the 1820s more than doubled the speed of the mule. As output rose, prices fell, making cotton goods accessible to an ever wider range of consumers. Britain found global markets: her expanding empire in India, which had once supplied cotton goods to the West, became a major market for British textiles.

Kashmir Paisley

Of the many Indian textiles exported to the West during the 18th century, woollens from Kashmir were the most successful. Handwoven, and thus expensive, the shawls were eventually copied by European manufacturers using Jacquard looms. Because Paisley in Scotland emerged as the leading center of such production, the shawl and its design became generally known as "Paisley."

Steam engines

An early Boulton & Watt beam engine from 1776, used by the Birmingham Canal Pumping Station, being dismantled in 1905 (above). The Boulton & Watt beam engine from *Rees's Cyclopedia* (left) and a later engine by Jonathan Hornblower, which used two cylinders in turn, from the same publication (right).

Water power

was still used where appropriate: the huge Laxey waterwheel in the Isle of Man was built in 1854 to pump water out of the local lead mines (right).

Steam on the farm

Portable steam engines such as those produced by Hensman and Sons were used to drive threshing machines and even to plough fields, using a system of cables to draw the plough. They were moved from field to field by horses. Stationary engines contributed to the expansion of industries based on agriculture, powering flour, sugar, oil and timber mills.

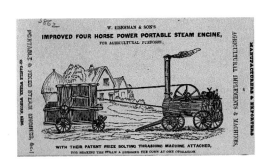

No other manufacturing industry was transformed by mechanization to anything like the same extent as cotton until later, but the use of machines to perform specific production tasks was becoming more widespread. Not long after steam engines began to be used to operate production machinery, steam-powered railway locomotives and ships revolutionized transport on land and sea. In agriculture, technological progress was much more modest. Even so, by the end of the 18th century, seed drills, threshers, which separated grain from straw, and the cotton gin, which removed seed from fibres, were being developed. Another half-century on, steam ploughs and reaping machines were becoming common. These offered substantial savings in the amount of human labor required.

Reactions to the factory

Apart from the textile industry, most machines – and workers – were located in small workshops, but the factory soon came to symbolize the machine age. To some this represented progress. For the advocates of industry, the factory was a hive of activity which could only improve the lot of society. In 1835, the chemist Andrew Ure likened the factory system to a "vast automaton" in which individual elements acted in "uninterrupted concert" under the guidance of a central force. Others were less enthusiastic. For Thomas Carlyle, the "mechanical age" entailed spiritual, artistic and moral poverty. Practical critics voiced more immediate concerns. To Friedrich Engels, the Manchester textile merchant and patron of Karl Marx, the relentless "tedium" of work in the textile mills stunted "body and mind," and contributed to social tensions in the manufacturing districts. To Richard Oastler, a Tory paternalist in Yorkshire, children in the factories were treated worse than slaves in the Caribbean.

Factories brought new forms of work and discipline. They cut across long-standing traditions and aroused the opposition of workers as well as social reformers. Factories did not "invent" long hours or child labor. Exploitation was common long before the Industrial Revolution. But factories did create an alien environment in which poor working conditions seemed less tolerable. The working day was no longer governed by the seasons or weather but by the factory clock, and the pace of work was dictated by machines. Strict discipline was enforced on younger workers by factory foremen rather than parents. Some employers sought to create benign regimes, others forced longer hours out of the workers to make their investment in machinery more profitable. Whether in factories or not, workers often had to put up with unsanitary or dangerous conditions. Dust was a major problem in textile mills, coal mines and the cutlery

The ideal

Andrew Ure saw the perfect factory as an automaton, in which humans were necessary only to insure the smooth running of the machinery. He used this illustration of a swan automaton in his *Dictionary of Art, Manufactures and Mines*, 1839 (left). Cotton mills came closest to this ideal, with machinery laid out along rational lines and a central power source, as in Jedediah Strutt's mill with its huge water wheel at Belper in Derbyshire (below). Few early factories looked like this: outside the textile industry, the transition from workshop to true factory was slow.

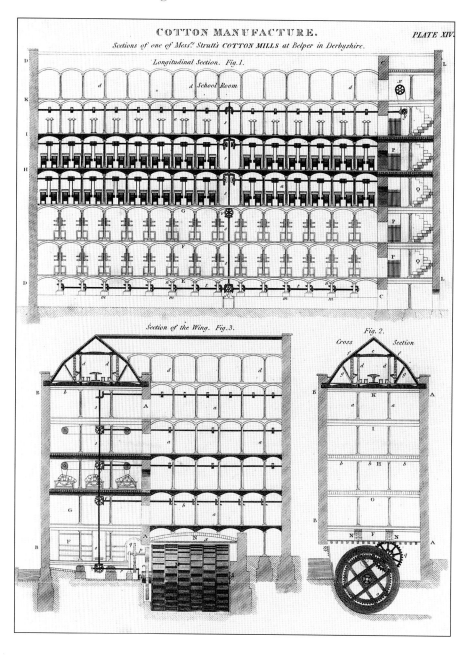

COTTON MANUFACTURE.   PLATE XIV.
Sections of one of Mess.rs Strutt's COTTON MILLS at Belper in Derbyshire.

Longitudinal Section. Fig. 1.

School Room

Section of the Wing. Fig. 3.

Fig. 2.
Cross Section

The reality

An engraving of William Darton's mill at Holborn Hill in 1820 (right). Even this somewhat sanitized view of child workers, probably winding warp threads, seems sinister to the modern eye.

The hazards of work

Many occupations required working in chronically unhealthy conditions, while others also involved acute danger: 350 miners were killed by an explosion at a colliery in Barnsley in northern England in 1866 (below).

trades. Working with unfenced machinery and hazardous materials frequently led to accidents.

With governments unwilling to interfere in industry and industrialists opposed to any restraints, statutory protection was generally slow in coming. In Britain, successful legislation in 1833 and 1844 went some way towards improving the lot of children and women, but working hours remained long and conditions hard throughout the 19th century. The same pattern of early abuses and long struggles for the protection of labor was repeated in all industrializing countries. Workers' own efforts to improve their situation met with mixed results. Skilled artisans in workshops were better organized and more successful than factory workers, but effective trade unions did not develop until well after 1851.

Another response, more reminiscent of the pre-industrial world, was violence. In 1811, framework knitters in the English Midlands destroyed a number of machines. These "Luddites" are often seen as the classic example of hostility to new technology, but their stance owed much more to earlier traditions of protest. They targeted the assets of bad employers rather than specific types of machine. In contrast, cotton workers in Lancashire frequently damaged machines seen as threatening their livelihoods. Similarly, the English agricultural laborers who took part in the "Captain Swing" riots of 1830 set out to damage the threshing machines which reduced the amount of work in winter. The actions of rioting cotton workers in Lodz in Poland in

Labor unrest

in the new industries often erupted into violence, and riots were often met with force. Two cotton-mill workers were killed during riots in Preston in Lancashire in 1842. Here, soldiers, backed up by top-hatted policemen, fire on the mob after the mounted magistrates have read the Riot Act.

King Ludd

A caricature of the unidentified leader of the machine-breaking riots, disguised as a woman. The term "Luddites" has been used to describe individuals or groups hostile to technological change, but the original followers of King Ludd were responding to a much wider range of grievances.

1861 were more deliberate. They destroyed the power looms, which had ruined the local handloom weavers, but ignored the spinning machines, which did not compete with local labor. Weavers in Prussian Silesia and silk workers in Lyons in France staged full-scale uprisings during the 1830s and 1840s. Perhaps the most tragic response to change was that of the handloom weavers in Britain, who vainly attempted to compete with the falling costs of mechanized weaving. The unequal struggle, sustained only by working longer hours for starvation wages, took decades to reach its inevitable conclusion. Although machines spelled doom for some occupations, they created demand for a wide range of new skills. Mechanics soon became an expanding and privileged group within the industrial labor force.

Machine tools and British toolmakers

As the use of machines grew and the range of their uses widened, so did the sophistication of the machine makers. In their pursuit of precision and consistency they developed better machine tools, the "machines that made machines" – lathes, screw-cutting engines, and planing machines. Their success enabled innovators to realize new designs. James Watt required a good seal for his steam engine cylinder and this demanded a higher level of precision than could be obtained using existing methods. We know that in about 1770 he ordered an experimental cylinder, "the best Carron [ironworks] could make," but was disappointed with it. However, by the time he reached Birmingham in 1774 to work with his new partner Matthew Boulton, he had the benefit of a new machine tool developed by the ironmaster John Wilkinson. This "boring mill" made Watt's invention a viable proposition. In the early 19th century, a wide range of machine tools was developed, allowing the manufacture of more accurate and complex metal or wooden components. These in turn aided the construction of higher performance machines.

The steam hammer

Various inventors had sought to apply steam power to the forge hammer, but the first successful design was that of James Nasmyth, with modifications by his works manager John Wilson. Nasmyth's hammer of 1840 (below) was a direct response to the increased demands of shipbuilding, as conventional hammers were unable to forge the huge paddle shaft planned for Brunel's *Great Britain*. Although the projected paddle shaft was abandoned when Brunel opted for screw propulsion, the steam hammer found a wide range of uses in engineering.

The planing machine

A machine for cutting a flat surface, Whitworth's planer of 1842 was a significant improvement on all previous designs (right). It was power-driven, offered greater standards of precision and was easier to work.

The Difference Engine

Charles Babbage's machine was designed to calculate and print mathematical tables automatically, thus reducing the possibility of human error. Only this small part of the engine was assembled (right), and the project was abandoned in 1833. His proposed Analytical Engine incorporated many of the features of the modern computer.

The screw-cutting lathe

Henry Maudslay, the most significant of all the machine toolmakers, adapted the slide rest for use with the general engineering lathe. The fixed cutting tool (below) could slide along the piece being worked as required, reducing the possibilities of human error and allowing a far greater range of machine parts to be cut more accurately.

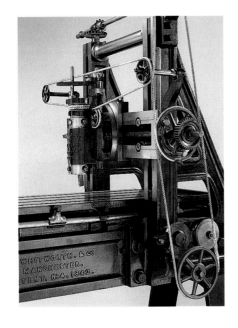

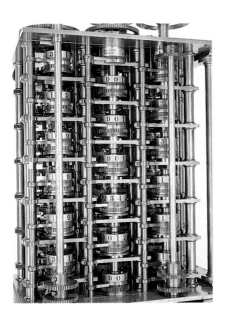

Many of these decisive contributions were made by talented engineer-inventors who acquired their skills in a relatively small number of workshops, notably that of Henry Maudslay. He became famous primarily for developing the screw-cutting lathe as an engineering tool, but was responsible for improvements to a wide range of other tools. Maudslay's insistence on accuracy was echoed by Joseph Whitworth, a former employee who later became a leading machine maker in his own right. Whitworth campaigned relentlessly for higher standards of precision and the standardization of measurements. James Nasmyth, the inventor of the shaping machine and improver of the steam hammer, had worked for Maudslay, as had Richard Roberts, whose inventions include the self-acting mule, a drilling machine based on punch-card technology, and numerous improvements to textile machines and machine tools. Joseph Clement had also worked with Maudslay and later became a contractor to Charles Babbage, whose Difference Engine is considered to be the first successful design for an automatic calculating machine.

Interchangeability and the "American System"

British machine makers contributed much to the country's reputation as the "workshop of the world" but, by the middle of the 19th century, it was becoming clear that developments across the Atlantic were more significant. The British government became increasingly interested in evaluating the potential of what came to be known as the "American System of Manufactures" and it was US rather than British practice which was to point the way to the later world of mass production and flexible specialization.

Despite their attention to accuracy and constant improvement of methods, British machine makers had not escaped – indeed, had not sought to escape – the traditional craft-based approach. They made large numbers of machines containing

Block making

The rigging of sailing ships required a huge number of wooden pulley blocks (above). The Portsmouth blockmaking machinery met the requirements of the Royal Navy, the world's biggest customer. One of the Portsmouth machines (left) and a pulley block (far left). The *Fighting Téméraire*, immortalized in Turner's painting of its inglorious last journey (opposite), was typical of the Royal Navy warships served by the Portsmouth factory. This 98-gun ship of the line, present at Trafalgar, would have required nearly a thousand pulley blocks. The painting has often been held up as a defining image of the triumph of the steam-driven industrial age, though it is also suffused with nostalgia for the glory of sail. It is ironic, therefore, that the *Téméraire* probably owed a vital element of its rigging to the beginnings of mass production.

components processed by advanced machine tools. The components, however, were not standardized enough to allow interchangeability. Every major machine was an entity in itself, lovingly finished by a team of highly skilled fitters. However well made, several individual components of such machines invariably required hand-finishing to achieve the perfect fit. Overall quality was dependent on time-consuming and expensive skilled labor. This raised the costs of production and of maintenance, as any spare parts needed later also had to be fitted by hand.

The concept of interchangeability of parts – where near-identical components could be used in machines of the same type without the need for additional fitting – was initially mooted in France, when the desirability of equipping large armies with standardized weapons was proposed by General de Gribeauval in 1765. An attempt to produce muskets with interchangeable components foundered in the face of the considerable technical problems involved. Complete standardization would have required carrying out large numbers of sequential operations to a degree of accuracy which even the most advanced machine tools had not yet attained.

The first successful foray into mass-production methods came in Britain. Each year the Royal Navy required large numbers of identical wooden pulley blocks for the rigging on its warships and these were obtained from specialist craft-based manufacturers. Marc Brunel, a refugee from revolutionary France and father of I. K. Brunel, had the idea that this complex manufacturing task could be mechanized. He was fortunate to meet the machine builder Henry Maudslay in 1802 and, after several years of collaboration, they produced what amounted to a dedicated production system, consisting of 45 special-purpose machines of 22 different types. Each block was moved from machine to machine to undergo an identical sequence of individual shaping operations with the minimum amount of human effort. Within a few years of the Battle of Trafalgar (1805) the system was in full swing, using ten skilled workers to produce 130,000 blocks per year – a volume of production which would have required 110 skilled workers using traditional methods. Although the Portsmouth blockmaking works was much visited, it failed to generate any significant imitative developments within British manufacturing. Why this should have been so remains an absorbing question.

Real progress was to come in the United States, where French military thinking formed a major influence. The concept of interchangeability of small arms had great appeal to the War Department, which fully appreciated the desirability of standardized weapons and was prepared to invest large sums in pursuing this goal. To a great extent, the federal armament factories at Springfield and Harpers Ferry acted as research laboratories, and private contractors were also offered generous terms to help them meet official requirements.

Change was still slow in coming. Even with a comparatively simple mechanism such as a gun, the large number of parts involved proved a far bigger challenge than the manufacture of identical wooden blocks. Nevertheless, with continued government support, fundamental improvements were gradually adopted. Among advances at the Springfield Armory was the use of stricter gauging techniques, where all components were measured against a master gauge. Meanwhile, at a private works at Harpers Ferry, John Hall used no fewer than 63 separate gauges to monitor the accuracy of components, and invented the concept of the "bearing point" – a single reference point on the workpiece. By designing all fixtures relative to the bearing point, Hall removed a major source of inaccuracies. American inventors devised new machine tools such as milling and grinding machines and the turret lathe, in which a succession of tools could be swung into place to carry out different operations on the workpiece. This became the mainstay of engineering production for decades to come. Also at Springfield, Thomas Blanchard assembled gun-stock making machinery

Nash lathe

Purpose-built barrel-turning lathe developed by the mechanic Sylvester Nash at the Springfield Armory in Massachusetts in 1817 (below). It was designed to produce the taper of a gun barrel but could not finish the flat oval shape at the butt end. In the Springfield area, five different barrel-turning lathes were developed at about the same time, and that of Thomas Blanchard would constitute the radical breakthrough.

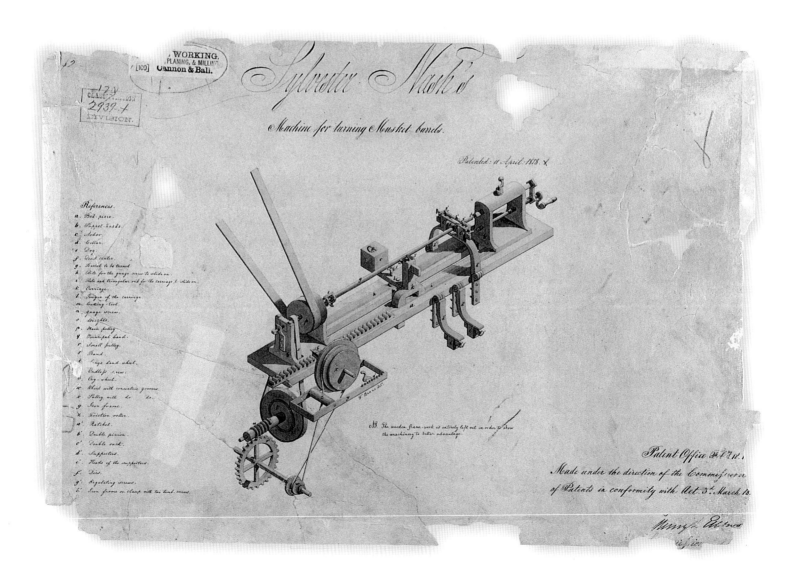

The Ames lock-recessing machine

1857, one of the American machines imported into Britain to re-equip the Royal Small-Arms Factory at Enfield (above). It was crucial that the barrel, lock and trigger assembly of the rifle made at Enfield should all be fixed to the wooden stock in the right relationship to each other. The purpose of this machine was to form the recess in the stock into which the lock would fit.

which became central to a sequence of 14 machines virtually eliminating manual labor from much of the process of gun making.

The path from "armory practice" to mass production was anything but straightforward. On purely economic grounds, the results were still far from encouraging. Even in the 1850s, the cost of producing guns with interchangeable parts was still higher than using traditional methods. In addition, the methods achieved did not yet translate into mass production – output was still relatively small, and firms which were able to increase production usually did so at the cost of interchangeability. Nevertheless, important lessons were being learned, and the ethos of interchangeability was being diffused elsewhere, as a stream of mechanics and engineers with armory experience left to found their own manufacturing ventures.

The multiplication of human skill through toolmaking and use of accurate gauges for verification was a central achievement and transformed working practice throughout the industrialized world. A cadre of "toolmakers" emerged with its own self-confidence – even arrogance – and these men were at the heart of production in all great factories.

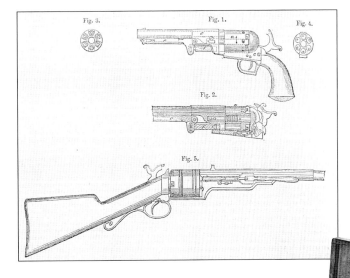

Fig. 3.   Fig. 1.   Fig. 4.

Fig. 2.

Fig. 5.

A London-made Colt (below).

Colt small arms

Samuel Colt's repeating revolvers and rifles (above) were adopted by the US Army. One of the sensations of the 1851 Great Exhibition, they helped stimulate British interest in the "American System of Manufactures," leading to the setting up of a Colt factory in London, where the latest American machine tools were used.

It was a town of machinery and tall chimneys… It had a black canal in it, and a river that ran purple with ill-smelling dye, and vast piles of buildings full of windows where there was a rattling and a trembling all day long, and where the piston of the steam-engine worked monotonously up and down like the head of an elephant in a state of melancholy madness.

CHARLES DICKENS
Hard Times, 1854

3 The industrial city

In the early part of the 19th century there was an unprecedented surge in population growth. By 1900 the population of Europe overall had doubled and that of Britain had tripled. New patterns of work brought more workers into factories, and new forms of infrastructure were introduced, including railways, water, sewerage, piped gas and later electricity. The "push" of inadequate rural employment combined with the "pull" of urban work opportunities to swell the cities. It was possible for a single female textile worker in the town to earn as much as the head of a rural family. This was not a phenomenon only of the most industrialized countries: migration from the Hungarian countryside made Budapest

Bricks.

An early 19th century hand-made brick from Lindsey, Lincolnshire (above), and a model of Henry Clayton's wire-cutting brickmaking machine from 1860 (right). By the middle of the 19th century, 500 million bricks were being produced every year in and around London.

Urban splendor

in 19th century Europe. The Place de la Concorde in Paris from across the Seine (left). The Luxor Obelisk at its center, a gift from Mehemet Ali, the Viceroy of Egypt, was erected in 1836, close to where the guillotine had stood during the French Revolution. He gave similar antique obelisks to Britain (Cleopatra's Needle) and to America. Budapest seen from Buda Palace Hill (below). The Széchenyi Chain Bridge across the Danube, designed by the British engineer William Tierney Clark, was the longest in the world when completed in 1849.

Europe's fastest growing city from 1867 to 1914. In 1851, 48 per cent of the population of England and Wales lived in towns, rising to 77 per cent in 1901. In the United States, only 4 per cent lived in towns in 1800, but 46 per cent a century later. The population of London, 1.1 million in 1801, was six times as large a century later, and New York had become a city of nearly 5 million people.

Contemporary observers were amazed by the growth of towns. Pessimists saw them as objects of fear – assemblages of the "dangerous classes" where disease, filth, ignorance and poverty were endemic – while optimists marvelled at the advantages of urban life. In the middle decades of the century, the citizens of many industrial towns built great town halls, often in a classical style, as expressions of civic pride and regional rivalry. In such centers as Berlin, Chicago and Manchester, grandiose universities were built throughout the century, symbolizing

the claims of new cities to represent a new culture. Many towns boasted Literary and Philosophical Societies and Mechanics' Institutes devoted to providing "rational amusement" to refine the tastes and soften the manners of city dwellers of all classes. Scientific subjects, including chemistry, optics and phrenology, were popular constituents of the many lecture courses run in such institutions. Phrenology mapped different mental faculties into specific areas of the brain, palpable as bumps on the surface of the skull. After a brief period of acceptance by medical men, it was mainly used by ordinary people to guide their lives in the disorienting world of the industrial centers.

Phrenological lectures
Mr. Droffnore was one of over 200 people who lectured on phrenology in Britain in the first half of the 19th century (left). Many businesses sold plaster casts of actual heads, often as sets of "felons" and "worthies," for use in lectures at Mechanics' and town institutes (below); these come from the Wadebridge Institute in Cornwall.

Civic pride

Splendid municipal buildings were constructed to foster a sense of identity in the rapidly expanding cities: town halls in Berlin, completed in 1869 (left below), Sheffield in Yorkshire, opened in 1897 (right) and the New State House, Boston, 1795 (above).

Midland Grand Hotel

opened in 1873 as part of the Midland Railway's St Pancras Terminal, London (left). Like many corporations, the MR was eager to announce its presence with a grand edifice, particularly as it was late in opening a London station. The architect Sir George Gilbert Scott, acknowledging criticism of his Gothic facade, declared the hotel "possibly too good for its purpose," but in self defense he could have pointed to the cast-iron beams he left exposed on its grand staircase.

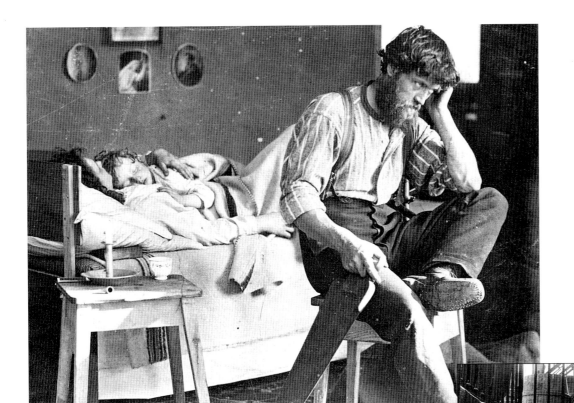

"Hard Times"

A rather staged depiction by the émigré photographer Oscar Gustav Rejlander in New York c.1860 (left).

Grinding poverty

was the lot of many city dwellers. Street urchins sleeping rough in New York in the 1890s (below).

Pollution

from industrial plants, sewage and garbage generated by the growing numbers of people created a logistical nightmare in cities lacking both the means and the laws to dispose of waste safely. Smoke from an early Manchester mill (left); the Great Dust Heap at King's Cross in London, a huge pile of ashes from a nearby brickworks which accumulated for decades before finally being removed in the 1820s (below).

Urban squalor

From the late 18th century the squalid domestic and working lives of the denizens of the industrial towns troubled the imaginations of novelists, social commentators and doctors. Poverty and squalor may well have been commonplace in the villages which many of the city-dwellers had left, but writers, including Henry Mayhew and Charles Dickens, focused on the human degradation of the new cities. In *Hard Times* (1854), Dickens portrayed the fictitious Coketown as a soulless machine-governed place, dominated by textile mills "where there was a rattling and a trembling all day long, and where the piston of the steam-engine worked monotonously up and down like the head of an elephant in a state of melancholy madness." For de Tocqueville, Manchester in 1835 was a "new Hades," with polluted air and water, rubbish piled high in the streets, and people crammed into cellars described as "repulsive holes," A decade later, Friedrich Engels made this Manchester squalor the subject of his *Condition of the Working Class in England*, published in his native German.

The census

A page from the 1841 census of Britain (left), in which the country's 18.5 million residents were obliged to give personal details such as name, occupation and approximate age. Later censuses demanded more specific information such as precise age and place of birth. Although bitterly resented by many as an invasion of privacy, census data provided the first complete picture of Britain's population, and proved vital for planning at both the local and the national level.

Ill health seemed a compelling index of the dangerous anarchy of urban growth. Statistical analysis of the health of towns had already become established at a national level in Britain, with the General Register Office (GRO) established in 1837 to collect information on births and deaths. From the time of William Farr's preface to the first GRO *Annual Report* in 1839, this data was analyzed to show geographical differences in causes of death. The first census was carried out in 1801, and every decade thereafter. New calculating technologies were used to process the masses of data and identify urban problems. In 1857 GRO staff acquired a Scheutz difference engine, a second-generation offspring of Charles Babbage's machine for tabulating mortality figures. It was employed in the compilation of the 1864 English Life Table, used in life assurance. In 1870 the Office purchased an arithmometer, an early type of mechanical calculator, to analyze the census. Before the end of the century, staff were also using slide rules. In 1890 the US census went further, using Hollerith punched-card machines to analyze increased quantities of census data; Britain followed suit in 1911. Some contemporaries, however, including Dickens, saw this stress on "fact" as symptomatic of the problems of industrial culture, placing emphasis on "the laws which govern lives in the aggregate" rather than the living conditions of individuals.

Cholera, smallpox, tuberculosis, dysentery, typhus and typhoid were among the many diseases suffered by the populations of the industrial cities. Asiatic cholera, which had broken out in India in

The Difference Engine No 3
constructed by the Swedish father and son Georg and Edvard Scheutz on the principle of the Babbage engine, the actual machine used by the General Register Office (below). It sought to eliminate errors of calculation and transcription by printing multiple copies of tables using papier-mâché plates.

CARLTON HILL EAST—continued.
22 Alston Mrs
20 Esse Charles Frederick, esq
16 Careless Miss
14 Chapple John, esq
12 Tiernan Edward, esq
10 Rawlings James, esq
6 Tasker William, esq
8 Wake William, esq
4 Anderson James Robertson, esq
2 Knaggs Sydney Henry, surgeon

Carlton hill vlls. Camden rd. (N.)
See Camden road, Holloway.

Carlton house terrace (S.W.),
bottom of Waterloo place.
1 TomlineGeo.esq.M.P.*for Shrewsbury*
2 Foljambe George S. esq. &Viscountess
Milton
3 De Clifford Baroness
3 Russell Hon. Mrs
3 Russell Hon. Edward Southwell
4 De Vesci Viscount
5 Bath Marquis of
6 Leinster Duke of, P.C
6 Kildare Marquis of
7 HardyJohn, esq. M.P.*for Dartmouth*
8 FitzGerald Lord & Lady Otho
9 PRUSSIAN EMBASSY,
Count von Bernstorff, ambassador
Frederick von Katte, sec. of embassy
M. Maurice Alberts } chanceliers
M. Paul Roux }
Rev. Adolphus Walbaum, chaplain
10 Ridley Sir Matthew White, bart. M.P.
for North Northumberland
11 Gladstone Rt. Hon. Wm. Ewart

16 Upward Mrs
18 Irving Rev. Joseph, M.A
20 Andrewes Thomas, esq
22 Brettingham Richard, esq
24 Smith George, esq
26 Pember Arthur, esq
28 Moore Francis, esq
HainesEdwd.Waltr.esq.(1Somrst.ter)

Carlton rd. Mile end and Old twn.(N.E.)
from Devonshire street, Mile end.
1 Russell George, bricklayer
6 Parr John, grocer
11 Baldwin Frederick, esq
11 Bridle John, timber dealer
15 StephensFleetwood Pembroke, grocer
20 Goodman Charles, builder
Stevens John, inspector of nuisances
(7 Carlton square)

Carlton st. Kentish New town(N.W.),
Wellington road.
4 Barker Mr. Robert
6 Marchmont John, esq
8 Whiddon James, academy
10 Physick Robert, sculptor
11 Fossett Mrs. Selina, seminary
71 Earl Thomas, artist
153 Snell James, dairyman

Carlton street, 12 Regent st.(S.W.)
4 Battcock Geo. wine & spirit merchant
5 Dyer Frederick, chemist
6 Trueman Henry, bootmaker
7 Maxsted Walter, hat & army cap mkr
8A, Weet Alexander, tailor
8 Chandler David, boot & shoe maker
10 Robertson Alexndr. Munroe, bootmkr
11 Alexander George, F.S.A. architect

28 Mostyn Rev. George Thornton, M.A
30 Hanbury John Capel, esq
31 Coon James, esq
32 Tewart Robert, esq
33 Messer Josiah, esq
34 Harper William Morris, esq
36 Dickins Henry Francis, esq
37 Clarke Adam, esq

Carnaby street, Golden sq. (W.)
19 Silver street.
2 Hoare Richard, hat maker
3 Taylor James, beer retailer
4 Keene Joseph, tailor
5 CrookEdwin & Fred.furnsh.ironmngrs
7 Reynolds & Abel, glass & china wa
8 & 48 Bywaters Geo.Hen.&Alfd. buildrs
8 Bennett Bros. manuftrng. goldsmiths
9 Webley John, coffee rooms
9 Gill Mrs. Matilda, dressmaker
10 Edwards Joseph, bootmaker
12 Gouldstone Uriah, coffee rooms
13 Ship, James Freeman
14 Watkins & Lane, grocers
15 & 16 Eyles Alfred George,news agent
17 Russell Mrs. Louisa, haberdasher
18 Platt John, corn chandler
19 Clark Henry William, greengrocer
20 Leigh JosephWilliam Flexney, oilman
...... here is Cross street.......
21 Shipton William Thomas Powley,
dining rooms
22 Zollor Francis, baker
23 Johnson John, grocer
24 Castle, Mrs. Maria Hempson
25 Earl James, carpenter
26 Taylor Thomas, dairyman
27 Hyde Thomas, grocer

A London directory

A typical page from the 1864 Post Office Directory of London (left), illustrating the wide social mix to be found in the big cities. While some streets consisted of aristocratic residences, others housed small businessmen or shops and workshops.

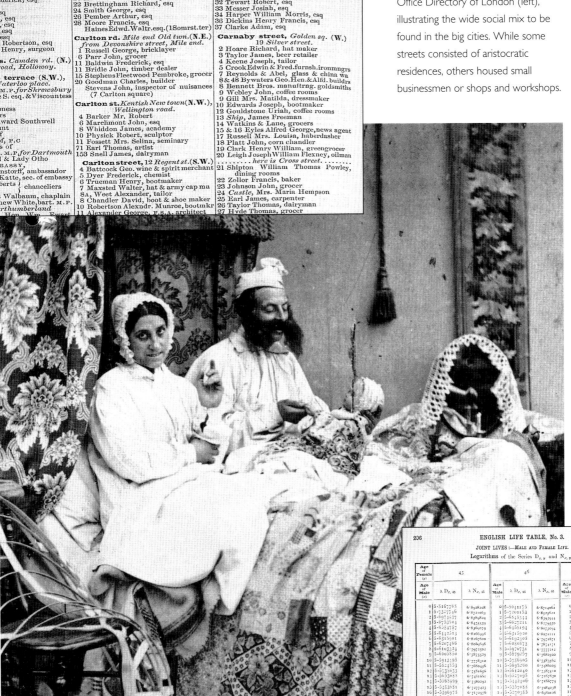

A domestic idyll

c. 1860. Such cosy, well upholstered – and carefully posed – surroundings (above) could be enjoyed by only a few. Note the rings of the wife's crinoline framework, to support her skirt, in the foreground: technology aiding fashion.

The English Life Table

(right) from 1864, compiled with the use of the Scheutz Engine (opposite).

Public disinfectors

in London, c.1877, employed by local authorities to administer disinfectant chemicals where infectious diseases appeared. While general standards of cleanliness were low, this was seen by many as a largely ineffective measure, dismissed as a "futile ceremony" by John Simon, Medical Officer of the General Board of Health.

1817, spread across Asia, North Africa, Russia and to Western Europe, reaching Britain and Ireland in 1831. A year later it had arrived in the Americas. In Britain alone, 30,000 died, although the total deaths from cholera across the century were not many compared to those caused by undifferentiated cases of "common fever." In 1854–5, in a third epidemic which bore particularly heavily on parts of London, the medical man John Snow calculated that the incidence of cholera was considerably higher amongst those who drank water pumped from a well in Broad Street, Soho. Removing the pump handle caused a slump in incidence. His separate study of cases amongst people served by different private water companies also provided evidence for his argument that cholera was a water-borne disease.

In the industrial cities, with the exception of London, deaths from smallpox declined across the century, although epidemics, particularly that of 1871–2, were felt across Britain. Alone among infectious diseases, smallpox had a specific preventive: vaccination, first publicized by Edward Jenner in 1798. Although adopted with alacrity in France where Napoleon introduced compulsory vaccination of the army, the practice provoked much resistance in Britain. In 1853 the British government passed the Vaccination Act making it compulsory for all

Cholera

A "Preventative Costume" from 1832 (above), and "The State Doctor" from c.1850 (left). Such satirical views of cholera reveal the variety of responses provoked by the epidemics. The doctor was one of a series which also featured a publican "fearing the spread of cholera" (presumably because of lost earnings) and an undertaker hoping to profit from it.

infants in the first three months of life. There was vigorous campaigning against the Act, on the grounds that compulsion was against the rights of the individual and vaccination against nature. In 1898 it was amended so that objectors were permitted to refuse vaccination of their infants.

Contemporary medical theory offered various explanations for the health problems of the urban working class. Doctors such as William Cullen and his pupil William Alison, for example, were trained in the new environmental medical theory of the 18th century. They saw the disproportionate incidence of fever among the poor as the product of debility caused by social and environmental factors, including diet, poor clothing, housing, lack of warmth, long hours of employment and the fumes and airlessness of many workplaces. But the breadth of this concern was displaced in the middle third of the 19th century by those who promoted political economy as the basis of "rational government." The single issue of filth, and the engineering required to remove it, took the place of the total environment as the explanation of urban ill-health. The miasmatic theory, which

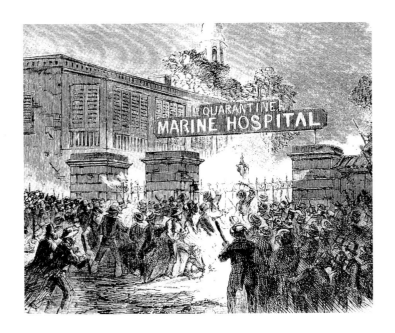

Quarantine

A mob attacking the quarantine station at the New York Marine Hospital in 1858. At the time, many people believed that such hospitals were responsible for epidemics.

Smallpox vaccination

as depicted by the French artist Louis-Léopold Boilly in 1807 (right). Once smallpox vaccination became compulsory, many different types of vaccination lancets were developed by doctors (above); the Mallam vaccinator from 1874 (above right) scratched the skin with gilded steel teeth.

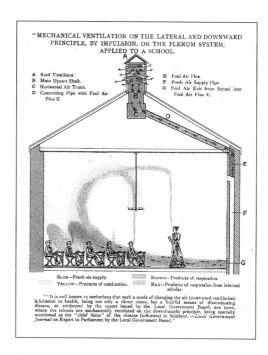

"MECHANICAL VENTILATION ON THE LATERAL AND DOWNWARD PRINCIPLE, BY IMPULSION, OR THE PLENUM SYSTEM, APPLIED TO A SCHOOL.

A Roof Ventilator.
B Main Upcast Shaft.
C Horizontal Air Trunk.
D Connecting Pipe with Foul Air Flue E.

E Foul Air Flue.
F Fresh Air Supply Pipe.
G Foul Air Exit from School into Foul Air Flue E.

BLUE—Fresh air supply.
YELLOW—Products of combustion.

BROWN—Products of respiration.
RED—Products of respiration from infected scholar.

"'It is well known to sanitarians that such a mode of changing the air (downward ventilation) is inimical to health, being not only a direct cause, but a fruitful means of disseminating disease, as evidenced by the report issued by the Local Government Board, one town, where the schools are mechanically ventilated on the downdraught principle, being specially mentioned as the "chief focus" of the disease (influenza) in Scotland.'—*Local Government Journal* on Report to Parliament by the Local Government Board."

explained disease in terms of an infectious "atmosphere" generated from decaying matter such as refuse and sewage, provided the theoretical framework for this approach for much of the 19th century. Germ theory came later (see page 68).

For the first three quarters of the 19th century, water was generally supplied to the towns by commercial undertakings, and sometimes several companies competed to provide water for the same district. It was not then commonplace for water to be piped into the individual dwellings of the poor, and all supplies were originally at low pressure and intermittent. Consumers would fill a cistern or bucket to last for several days until the supply was next turned on. The engineer Thomas

Ventilation

of the home, school and workplace was vigorously promoted in the closing decades of the 19th century, as (left) in the Boyle System of Ventilation (c.1880). Robert Boyle also published a series of pamphlets entitled *Sanitary Crusades*.

Water

The beam engine and boiler houses of the Nottingham Water Works, 1856 (below), a project of Thomas Hawksley, first an ally and then an opponent of Edwin Chadwick.

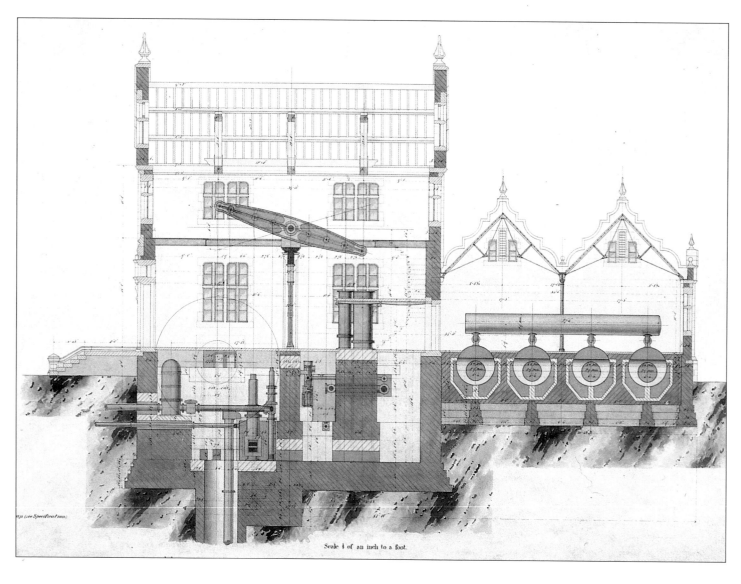

Scale ⅛ of an inch to a foot.

Education in hygiene

A lecture cabinet from 1895, made by Charles Campbell, member of the Royal Sanitary Institute. The cabinet contained miniature drainage and sewage fittings, sanitary appliances and ventilation equipment, for teaching the principles of hygiene to trainee medical officers of health.

Imperial Gas Works

at King's Cross, London. The retort house in the 1870s (above), with workers shovelling coal into the ovens. Gas for heating and lighting was produced, by heating coal in the absence of air, in a large number of such gas works. These distributed gas at low-pressure, locally, unlike today's high-pressure natural-gas networks which distribute gas over long distances. The residual carbon from the gas-making process was then sold as coke for fuel.

Hawksley applied his experience of gas supply, in which gas is conveyed under pressure in sealed pipework, to provide a constant water supply to 8,000 homes in Nottingham from 1830. He was one of many engineers who moved into sanitation, installing miles of underground pipes and building steam-powered pumping stations to feed water from great distances.

The provision of systems to remove human sewage and drain away rain water became invested with utopian values. Civil engineers may have been the first to install new sanitary infrastructures, but it was a British civil servant who exalted sanitation and made it the core ideology of public health. Edwin Chadwick, often seen as a reforming zealot, held a vision of an "arterio-venous" system in which clean water would be introduced into the towns to act as a vehicle to transport disease-causing excrement and other filth to the countryside, where it would be sold as manure, generating enough profit to pay for the construction of the sewers. He argued vehemently with engineers that glazed pipes should be used in preference to brick sewers, which they favored, to carry waste away from the cities. The smaller the pipes, he contended, the faster disease causing filth would be removed. However, civil engineers such as Joseph Bazalgette, who created the main London system and the associated

Joseph Bazalgette's
Thames Embankment, shown under construction in 1867 (left). Behind it ran the Metropolitan Railway and a sewer underground, with a new road above. It reclaimed 32 acres of mud from the Thames. The Northern Outfall sewer, near the Abbey Mills pumping station (below), was an integral part of Bazalgette's scheme; sewage gathered here from London was pumped eastwards for eventual discharge into the river well away from the city. In this photograph from 1862, Bazalgette can be seen standing, above the completed sewer tunnel, on the right.

Haussmann's Paris:
the Champs-Elysées (top); the city sewers, c.1860 (above).

Thames Embankment, did not agree with Chadwick's theories. Bazalgette's scheme involved the construction of 82 miles of brick sewers beneath London and used pumping stations to discharge the waste. When completed in 1865, the system was capable of dealing with 420 million gallons of sewage and rain water daily. Most of Bazalgette's system is still in use today. A "water mania" swept Europe, resulting in new sewer and water supply infrastructure in virtually every major northwestern European city by the close of the century. While Parisian streets doubled in length under the town planner Baron Haussmann, the sewers increased by more than five times. In the United States, the 16 largest cities all had waterworks by 1860. The other major practical measure taken by health reformers was the ventilation of buildings. Many citizens preferred to seal their homes against miasma and smoke, but numerous public buildings were fitted with roof-mounted ventilators and ground level inlets, which can still be seen today.

Housing and urban improvement

A new word, "slum," passed from slang to orthodox use during the first half of the century to describe poorly built and deteriorating areas of housing, often associated with poverty and disease. The pace of migration into the towns had brought heavy overcrowding which increased the risk of epidemics. Some employers, such as David Dale and later Robert Owen at New Lanark, Titus Salt at Saltaire, the Lever Brothers at Port Sunlight and Krupp at Essen provided model housing for their workers, but, in an age when virtually all domestic property was rented, many landlords saw little point in improving their properties. In Britain societies were formed to construct improved housing for the working classes. Later, from the 1860s, private bodies such as the Peabody Trust built blocks of model dwellings. More extensive improvements in housing came with the spread of active local government, equipped with greater powers, after World War I.

New York slums

in 1887 (above), photographed by Jacob Riis, a reporter who later became a social reformer. Whole families dwelt in such makeshift housing.

Model housing

New Lanark, where education and healthcare were provided for the cotton-mill workers in the opening decades of the 19th century (bottom left). Port Sunlight (right), built from 1887 onwards; a complex containing 800 houses, a library, museum, two schools and a technical institute for soap makers. In the closing years of the century Krupp erected quality housing for its steelworkers at Essen in Germany (below), but the influx of workers into the city was so rapid that fewer than 1 in 30 could be accommodated in such buildings.

Gas
Greenwich gas works on the site which now houses the Millennium Dome (left); the Phidomageireion, a gas cooker of 1853 (below left) promoted by Alexis Soyer, a famous French chef who reorganized catering in the Crimean War in 1855; the Welsbach type C gas mantle of 1893 (left), the first gas lamp to produce a really bright light – from the radiating mantle rather than the flame itself.

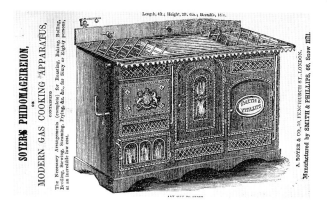

In Glasgow from 1866, local government intervention led to urban renewal. City services were improved, parks were provided and a municipal center and art gallery were built. Under the leadership of Joseph Chamberlain, using Glasgow's model, Birmingham embarked on a scheme of slum clearance and urban regeneration from 1875 onwards. Here and elsewhere in the industrialized world, areas were set aside as public parks, and amenities such as sanitation, gas lighting and paving were slowly extended to the poorer areas. Early in the century gas, manufactured from coal, had become available for illumination and heating, first in London and then rapidly throughout the newly industrializing world. Gasholders and street lamps were added to the urban skyline. These

Gas lighting

A street lamp in the Piazza di Monte Cavallo, Rome, in an 1841 daguerreotype by Alexander John Ellis (right). First used to light London streets in 1812, gas lamps appeared in all major industrial cities over the following decades. After dark, much of urban life became dependent on gas lamps both inside and on the streets: a miscellany of inconveniences arising from a gas workers' strike in London in 1872 (opposite, below).

The Underground

Chancellor of the Exchequer William Gladstone, his wife and other notables, at Edgware Road Station, London in May 1862: the first trial journey of the Metropolitan Railway, the world's first underground line, linking Paddington station in the west to the City in the east. They were hauled by steam engines which produced unpleasant fumes in the confined tunnels.

were only the most visible parts of the extensive systems which included gas-works, tanks and the piping which ran beneath city streets. By the late 1830s the 12 London gas companies had invested huge sums. At night the streets could become places of excitement rather than danger.

Throughout the century, the urban landscape was shaped and reshaped by new forms of transport. From 1830, railways linked larger towns, and the system was soon extended to serve the suburbs. Railway lines often cut through slum areas, displacing the poor, and influenced the pattern of subsequent urban growth. Horse-drawn streetcars, which appeared in America during the 1850s, were to be seen in European cities during the 1860s and 1870s. By 1900, electric streetcars were a feature of many major towns, permitting rapid transit to the suburbs.

New technologies enabled cities to spread upwards as well as outwards. The steel frame and the elevator allowed the

construction of taller buildings. In 1885, the era of the skyscraper arrived with the opening of the Home Insurance Building in Chicago. New utilities were added alongside water and gas pipes and sewers. From 1871, in London and other major cities, hydraulic power companies supplied water under high pressure through underground pipes to power not just elevators, machine tools and dock cranes but also the revolving stages of several theatres. A pneumatic tube

Skyscrapers

usually built to house large corporations, helped transform the centers of many American cities from the 1880s. The weight of the walls was carried by an inner steel skeleton rather than by masonry. At 302 feet, the 22-story Masonic Temple in Chicago (left) was one of the tallest in the world when completed in 1892. By 1913, New York could boast the 55-story Woolworth Tower. A building, made fireproof by tiles around the steel frame, under construction in Chicago in 1894 (above). Numerous other inventions, such as the elevator, central heating and electricity were also needed to make skyscrapers practical.

New York

The pneumatic tube mail transmitter at Brooklyn Post Office, c.1899 (right). About 500 letters could be dispatched at a time through the tube to New York General Post Office nearly two miles away. This system covered 27 miles and remained in use until 1953. The New York Elevated Railroad (the El) was one of the city's main forms of urban transport prior to the opening of the Subway in 1904. A station on the El, c.1874 (below); an El train winds its way through the city, c.1884 (bottom).

system delivered paper copies of telegrams between the Central Telegraph Office and a network of post offices from the 1850s onwards. During the 1880s and 1890s the London skyline was marked by large numbers of telephone posts and wires. The 1892 Telegraph Act, however, prevented a worsening of this problem by permitting companies to install cables underground.

New biological theories and the city

Towards the end of the century, two new forms of medical and biological theory began to make an impression on public health practice. From the 1870s, the miasmatic theory was partially overtaken by the greater explanatory power of bacteriology, championed by Louis Pasteur, Robert Koch and others. In the new theory specific germs were thought to cause specific diseases. This provided a new theoretical basis to understand and improve sanitation, but the new germ-centered public health was even less concerned with the broader environmental determinants of ill-health than Chadwickian sanitation had been. For several decades, the bacteriological laboratory, with its microscopes, petri-dishes and incubators for the cultivation of microbes, became emblematic of public health activity. Germ theory guided British legislation. Provision for isolation hospitals was enshrined in law in 1893, and in 1899 an Act for the notification of dangerous diseases was passed.

In this period social Darwinism also came into vogue. From the 1850s, commentators expressed fears that urban populations were degenerating, and that the physical and mental energies of city dwellers,

Louis Pasteur

at work on his early sterilizer, c.1870 (above). Pasteur proved that infectious diseases were caused by micro-organisms rather than by polluted air. He discovered that these germs could be killed by heat, and devised methods of sterilization. Pasteur's work caught the popular imagination, and he gained wide acclaim as a hero of science.

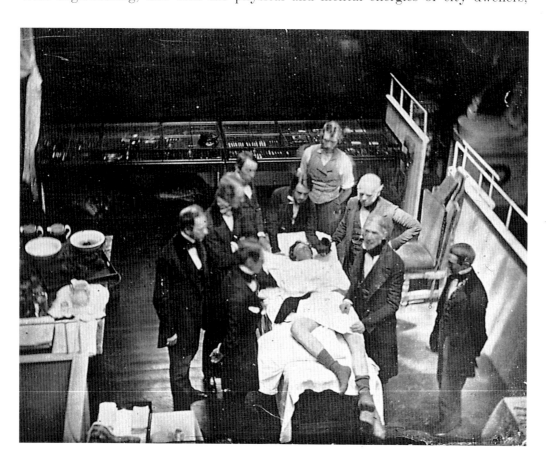

Anaesthetic

A surgical patient anaesthetised with ether at Boston's Massachusetts General Hospital in the late 1840s (left). The hospital had pioneered the use of ether in 1846, a year before the anaesthetic properties of chloroform were discovered by the Scottish physician James Young Simpson.

Tuberculosis

Robert Koch isolated the tubercle bacillus in 1882. The incidence of the disease decreased gradually from the mid-19th century where living conditions improved. Nevertheless, TB remained one of the most feared diseases in all sections of society until the development of effective treatments after World War II. A French poster (right) promotes a campaign against tuberculosis and infant mortality.

Serum therapy

Bottle with packet of the original tetanus serum, c.1900 (below). Sera effective against tetanus and diphtheria were developed by the German Emil von Behring and the Japanese Shibasaburo Kitasato, both assistants of Robert Koch in Berlin. These sera contained natural anti-bodies to the tetanus and diphtheria bacilli induced in animals by injection with weak doses of the diseases.

VISA Nº 15996 — DEVAMBEZ, PARIS.

LA VISITEUSE D'HYGIÈNE VOUS MONTRERA LE CHEMIN DE LA SANTÉ ELLE MÈNE UNE CROISADE CONTRE LA TUBERCULOSE ET LA MORTALITÉ INFANTILE. SOUTENEZ-LA !

COMMISSION AMÉRICAINE DE PRÉSERVATION CONTRE LA TUBERCULOSE EN FRANCE

especially the working classes, were being overtaxed. Urban overcrowding, poor ventilation and diet, exacerbated by alcohol and tobacco consumption would, it was feared, lead to weaknesses which would be inherited by the next generation, bringing about a terminal decline of the population into bestiality. The theoretical foundations of eugenics, popular among many reform-minded individuals until the 1930s, were elaborated by Darwin's cousin, Francis Galton, who built a structure of statistical method on such degenerationist fears.

Town and Country

Many of the problems experienced in Britain early in the 19th century were later endured in other countries, particularly where the pace of industrialization was vigorous. The barrack-like housing in Berlin and the Ruhr, and the squalid accommodation for workers in St Petersburg, demonstrated the difficulties of coping with rapid urbanization. Even in towns which were no longer expanding, the environment deteriorated quickly wherever poverty was prevalent. To those hostile to the changes associated with industrialization, the city stood for everything that was unhealthy in the modern world. After the 1880s, when faith in unlimited economic progress diminished, many commentators pondered what Thomas Carlyle had originally called the "condition of England question." It was not until 1892 that Engels' book (see p.53) was first published and widely

Central Park
in New York, c.1865 (above); the first landscaped public park in the United States, modelled on the public spaces of London and Paris. Opened in 1859, the park involved the transformation of over 800 acres of swampland, and the dispersal of shanty communities. About 20,000 workers moved nearly 3 million cubic yards of earth and planted over a quarter of a million trees. Mehemet Ali's obelisk was erected here (see p.49).

Slums
Typical working-class housing in 1890s London (left); the inhabitants shared water taps, and a gutter ran through the centre of the road.

Yosemite National Park

Placing areas of outstanding natural beauty under government protection was one practical response to urbanization and the disappearance of large tracts of wilderness or countryside. Yosemite, established in 1890, had its roots in the special nature reserve set up by Abraham Lincoln in 1864. A cart passing through a section cut out of the base of a giant sequoia in Mariposa Grove, Yosemite, c.1870 (below); photographing Yosemite Falls from Glacier Point, c.1880.

Garden cities

houses and gardens in Letchworth Garden City (below), a practical experiment in creating rural values within an urban setting.

read in England. The Booth and Rowntree surveys mapped the income levels of different households within London and York respectively. There were also journalistic works and novels including Andrew Mearns' *The Bitter Cry of Outcast London* and George Sims' *How the Poor Live* (both 1883), while Arthur Morrison and Jack London penned moving condemnations of the squalor in London's East End. In Russia and Poland, Mamin-Sibiryak and Reymont wrote about local conditions in terms that Engels might have used fifty years earlier.

One consequence of the anti-urban trend was a romanticization of the rural environment, portrayed as a healthy and clean alternative to the horrors of the city. The trend was strongest in Britain, the most urbanized country. From the beginning, this involved a return to an imaginary past. In Britain in 1846, the radical politician Feargus O'Connor launched his "Land Plan" to enable town dwellers to return to the village. Forty years later Jesse Collings, a mainstream politician, could still coin the slogan "three acres and a cow" to sum up what he believed to be the aspirations of many British town dwellers. The notion of the rural idyll persisted. Early in the 20th century it found concrete expression in Letchworth, the first of the Garden Cities, founded by Ebenezer Howard, and designed by Raymond Unwin.

The part which he [Watt] played in the mechanical application of the force of steam, can only be compared to that of Newton in astronomy, and of Shakespeare in poetry

E. M. BATAILLE,
Quoted in Samuel Smiles, *Lives of Boulton and Watt*, 1865

4 The age of the engineer

If the impact of the Industrial Revolution was patchy, the far-reaching nature of changes in technology became increasingly clear to many in the mid-19th century, as new transport and communications networks extended across countries and continents. Such innovations were brought to popular attention through deliberate publicity campaigns, and the promotion of progress also became institutionalized through the holding of regular exhibitions, showcases for new products and inventions. One outcome of the popularization of the notion of progress was the increasing adulation of the engineer, personified as the creative force responsible for the transformation of the landscape and the dramatic changes affecting so much of everyday life. Among the first to receive such treatment was James Watt, eulogized as

Robert Stephenson's *Rocket* of 1829, an experimental engine which firmly established the viability of the locomotive. It brought together into one design three crucial innovations: the multi-tubular boiler, the blast pipe, and direct drive from piston to wheel. None of these was original to the *Rocket*, but their use in combination was highly successful.

the "modern Archimedes" for single-handedly conceiving the steam engine, the mighty invention which formed the basis for Britain's greatness. As early as 1824 a monument to Watt was commissioned for Westminster Abbey, and before long, provincial towns were erecting statues and memorials to him as well. In 1853, the Royal Navy launched the 90-gun ship *James Watt*, thereby conferring an honor normally reserved for dead admirals, members of the royal family and classical deities.

The stereotype of the lone genius, usually from a humble background, struggling with adversity to become a benefactor of mankind, was parodied by Dickens in the 1850s in *Bleak House* and *Little Dorrit*. In the following decade, Samuel Smiles established a whole pantheon of heroes of the Industrial Revolution in his *Lives of the Engineers* and *Industrial Biography*. These depicted their subjects as paragons of self-help, combining mechanical genius with infinite patience and industriousness, and went a long way

The Watt Memorial
in Westminster Abbey (far left) by Sir Francis Chantrey. The inscription hails the inventor as one of the "real benefactors of the world." He sits with his dividers in his right hand.

William Murdoch
A bust (left), also by Chantrey, of this employee of Boulton & Watt, who won fame as the pioneer of gas lighting, though his many vital contributions to Watt's engines received little recognition.

The heroic vision

the young Watt demonstrating his fascination with steam (opposite above); the mature Watt as the lone genius – James Eckford-Lauder's *James Watt and the Steam Engine: the Dawn of the 19th Century*, painted c.1855 (above).

towards establishing an authorized version of engineering progress, leading to perceptions which have never quite been eradicated.

The popular notion of single-minded "great men" often failed to do justice to their versatility. Although they were exclusively identified with specific inventions or undertakings, their interests were invariably broader. Watt had been a surveyor and instrument maker before developing an interest in steam engines. Joseph Bramah, best known for the machine tools he invented and as an hydraulic engineer, also invented a flushing toilet and an "ever-pointed" pencil. William Armstrong, another excellent hydraulic engineer and entrepreneur, also conducted experiments with electricity in his spare time.

Many heroic figures were motivated by far less exalted ideals than the mythmakers chose to assign to them. Boulton and Watt's aggressive defense of their patent rights delayed further improvements to the steam engine in Britain by several years, and George Stephenson's quarrelsome nature

Neglected legacy

an adapted atmospheric steam engine at Coalbrookdale (left). Devised in 1712, such machines were often used to pump water out of coal and tin mines. Cheap and easy to maintain, they remained in use long after the advent of the efficient but expensive Watt engine. The machine in the picture, installed as late as 1790, had probably been built on the spot.

insured that his contribution to early railways was not always positive. Moreover, the pantheon of great men was unfairly selective. In many cases, individual heroes were given (or assumed) the credit for the actions of others. Thus George Stephenson was happy to take – and continues to receive – credit for much of the work of his son Robert. Earlier, James Watt had taken great pains to discredit any other claims to innovations in steam-engine technology.

Even if inventors did make decisive contributions, they invariably borrowed freely from the work of predecessors, or colleagues. The extravagant praise heaped on Watt ignored the earlier Newcomen engine and the work of contemporaries such as Trevithick or Hornblower. James Nasmyth did not single-handedly introduce the steam hammer. Alexander Graham Bell's contribution to the telephone extended little beyond the first imperfect prototype. The persistent emphasis on a small number of well known inventions plays down the achievements of a legion of unjustly forgotten technicians responsible for the continuous stream of steady improvements which insured that technology never remained static.

The public could be extremely fickle. Some inventors, skilled in the art of self-publicity, could receive full recognition for their efforts, others could be ignored, or receive only belated acknowledgment. Samuel Morse eventually achieved legendary status in the United States – a statue to him was unveiled in New York's Central Park during his lifetime – but only after years of similar treatment in Europe, where he had long been showered with decorations and honorary titles.

Oliver Evans

the American inventor who constructed a high-pressure steam engine in 1804. James Watt had dismissed the use of high pressure as dangerous and discouraged further research, but the work of Evans, Trevithick and others made it clear that the future of steam power lay in this direction.

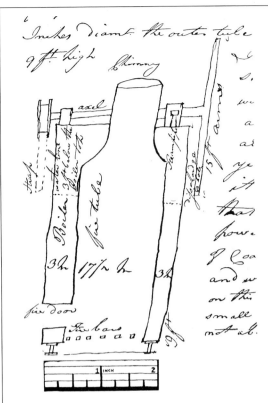

Richard Trevithick

His sketch for a recoil engine and boiler (left), a type of reaction turbine. The arm on the right was designed to rotate like a catherine wheel. A poster for his portable engine, c.1800 (right); his London Railway of 1808 (below), a brief form of public entertainment in the days before the full potential of his engine was understood.

Stephenson and Brunel

Even if Smiles and his contemporaries painted an idealized picture of certain engineers, the profession as a whole earned the accolades heaped upon it. The examples of Robert Stephenson and Isambard Kingdom Brunel show what seemingly limitless capacity for hard work allied with brilliance and versatility could achieve. Both died within weeks of each other in 1859, aged 56 and 53 respectively, outliving their famous fathers by only a decade. Both had been worn out prematurely by a succession of demanding projects, invariably at the cutting edge of current practice.

Robert Stephenson had begun an apprenticeship as a mining engineer, but his career was almost entirely devoted to the railways. Apart from designing the *Rocket*, which demonstrated once and for all the viability of the steam locomotive at the famous Rainhill Trials in 1829, he worked on a number of major schemes, from the pioneering Stockton & Darlington (1825) and Liverpool & Manchester (1830) lines which launched the railway era, to grander projects linking London with Birmingham and the north, as well as many lines in Europe.

North Church Tunnel
(below), one of nine on the Stephensons' London and Birmingham Railway. This line, opened in 1838, cost £53,000 a mile and Euston, its London terminus, was the capital's first big station. Robert Stephenson (left).

The Rainhill Trials
A later depiction of the triumph of the *Rocket* in 1829 (left).

The Britannia Bridge
(below) taking the London to Holyhead Railway across the Menai Straits; (inset) a cross-section of the tubular bridge; in Stephenson's innovative design, rigidity was maintained by the walls.

Railway building
involved much more than simply laying the track: a cutting excavated through rock at Olive Mount, part of the Liverpool and Manchester Railway, on which Robert Stephenson assisted his father George, (left).

Brunel was the general engineer par excellence, with spectacular contributions in several fields. He is remembered primarily for the Great Western Railway, always regarded as the greatest of the British lines, and for the glorious Clifton Suspension Bridge. He also collaborated with his father on the Rotherhithe Tunnel (opened 1843), the world's first tunnel under a river bed. Probably his greatest contribution to engineering was the construction of three mammoth steamships, the *Great Western* (1837), *Great Britain* (1845) and *Great Eastern* (1859), which marked successive milestones in the use of steam over long distances.

Brunel

(left) and with his customary cigar, at the first abortive attempt to launch the *Great Eastern* in 1857 (near right); work on the Royal Albert Bridge at Saltash, which took the Great Western Railway across the Tamar estuary into Cornwall. It was completed in 1859, months before Brunel's death (far right). *Rain, Steam and Speed, Great Western Railway*, Turner's evocative depiction of a locomotive in full steam (below).

Transport and communications

Textile machines, the stationary steam engine, and advances in metallurgy were not necessarily visible to the average inhabitant outside the main industrial towns, but the coming of the railway or the launching of a regular steamship line galvanized public opinion. If a small minority bemoaned the clearance of buildings or stretches of countryside in advance of a railway line, the majority was highly enthusiastic. Inclusion in a new transport route brought the promise of increased trade and prosperity; access to a network seemed virtually synonymous with membership of the modern world. Each stage of the process served as an occasion for celebration. One small French town was said to have rung its church bells simply on hearing the news that a proposed railway line passing through it had been duly authorized. Carefully preserved wheelbarrows and spades used to dig the first soil testify to the symbolic significance attached to the inauguration of railway construction works. Even when crowds were absent, symbolism was maintained for the camera: a golden spike was used to make the final connection of the first

Glass goblet
commemorating the opening of Robert Stephenson's Newcastle High Level Bridge in 1849 (above).

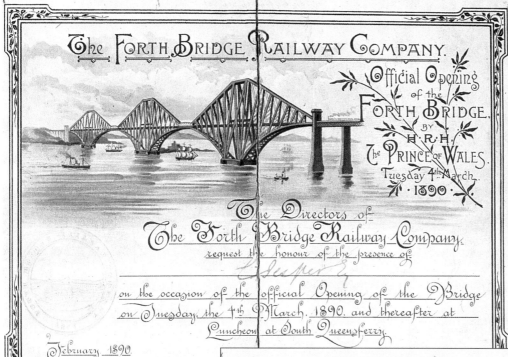

The first German line
The celebrations accompanying the opening of the Nuremberg to Fürth railway in 1835 (below).

An invitation card
to the official opening of the Forth Bridge in 1890 (above). The Prince of Wales, later King Edward VII, performed the ceremony by symbolically closing the last rivet with a silver key.

God Speed

Church and State often came together to endorse progress, as at the benediction of the Rouen and Le Havre Railway in 1847 (above).

transcontinental rail link across the United States in 1869. Most opening ceremonies involved festivities on a lavish scale, with the attendance of the company directors, engineers or contractors alongside representatives of the municipal authorities and local politicians and worthies. Celebrations frequently got out of hand – New York's City Hall almost burned down during the frenzied scenes following the linking of the British and American telegraph networks in 1858.

Once a railway had successfully linked two major cities – Liverpool and Manchester in 1830 – the way was open for the construction of large numbers of lines throughout Britain and northwestern Europe, forming the beginning of regular networks linked by telegraphs and run with the help of timetables, cost accounting and signalling systems. British contractors and workmen were heavily involved with the building of railways in Europe and elsewhere,

Euphoria in New York

The enthusiastic reaction to the first exchange of telegraph messages between Washington and London in 1858 (right). The revellers could not have foreseen that the cable would break within weeks.

but it was the sheer scale of the American lines which required the greatest engineering feats. By 1890 the United States had built a network larger than that of western Europe, extending over 150,000 miles of line. By the end of the century, railways had been built through some of the most remote regions, opening up the rich untapped resources of the New World and the areas of recent settlement in Australasia, South Africa and Latin America. In India, Central Asia and Africa, major lines were constructed to serve the interests of the great empires. Railways were to change everyday life in other ways even for those who never travelled – the establishment of networks necessitated the standardization of time, first within individual countries, and later across the whole world. In Britain, the Greenwich standard gained general acceptance mid-century and was legally confirmed in 1880.

Railway excursions

A poster advertising trips to New York to see the *Great Eastern* steamship (above).

The Transcontinental link

The meeting of the Union Pacific and Central Pacific railroads at Promontary in Utah in 1869 (left).

An American locomotive

crossing a wooden trestle bridge during the construction of the Transcontinental Railway (above).

Four years later, it was set as the zero point for international time.

Just as the railway had revolutionized land transport, the use of the steamship transformed maritime trade. Steady improvements to marine engines – mainly to reduce coal consumption – enabled steamships to extend their range and bring down costs. Although Brunel's great vessels had not been commercial successes, several regular transatlantic services were thriving in the 1860s. By the end of the century, sailing ships had been virtually

US military railway depot

at City Point, Virginia, in 1863 (right); the American Civil War was the first major conflict in which railways played a strategic role.

banished from the main trading routes. Freight charges and passenger fares dropped considerably. This proved a great stimulus to international trade and enabled large numbers of people to cross the oceans in search of a better life. Between 1880 and World War I, over thirty million Europeans moved to other continents. With such journeys now cheaper, faster and safer than ever before, not all those who travelled intended to settle – significant numbers of temporary migrants were returning to Europe with money earned overseas.

Other great advances came from the use of electricity to transmit information. Several early proposals for an electromagnetic telegraph came to nothing until the collaboration between Cooke and Wheatstone produced the first practical model in 1837. Telegraphs were developed in the United States by Samuel Morse and his partner Vaile, who adopted the dots and dashes code widely used until recently. National telegraph lines were laid alongside the railways. More challenging was the establishment of international links

The Thames Iron Works'
engineering and shipbuilding premises, London 1867 (left).

The Hamburg-Amerika line
carried huge numbers of European emigrants to the United States: the *Kaiserin Augusta Victoria* passing the Statue of Liberty (right) in 1890.

Calcutta c.1900
Steamships in the Hooghly River being loaded or unloaded. Their derricks swing the cargo into the holds or waiting lighters (below).

using undersea cables. London and Paris were connected in 1852 and the first transatlantic cable was laid in 1858. The early cables proved unreliable, and permanent connections took several years to establish, but by the 1870s Europe had links with India, the Far East and Australia.

Communications technology was further transformed with the conversion of the human voice into transmittable form, thanks to the development of the telephone in 1876. The first international line opened between Brussels and Paris in 1887, and by the end of the century nearly two million phones were in use. Many of the improvements to the telephone came from Thomas Edison, an excellent example of the engineer as scientist, inventor and entrepreneur combined. The work of pioneers like him and Sebastian de Ferranti paved the way towards the wider use of electricity and the establishment of a power supply network, which would eventually embrace every home and business in the developed world.

With the growing complexity of the various technological systems and the rapid expansion of the specific knowledge and experience required to remain at the cutting edge of developments, the all around brilliance of a Brunel was no longer enough, and the profession branched out into several specialist areas. It was soon possible to distinguish between civil, mechanical, electrical, sanitary and marine engineers, to name only the most important branches of the profession.

As industrialization spread, many countries affected soon acquired their

The *Great Eastern*

taking on board the third transatlantic telegraph cable in 1866 – the first to achieve a permanent link between Britain and North America (left); coiling the cable below deck (above); cross-sections of the cables of 1858, 1865 and 1866 (inset).

The Cooke and Wheatstone telegraph

The five needles moved so as to point at a succession of letters, thus building up the message (right). Samuel Morse (above right), inventor of the code that made this machine obsolete. In 1854 Charles Dickens remarked that "Few things that I saw... took my fancy so much as the electric telegraph piercing like a sunbeam, right through the cruel old heart of the Coliseum at Rome."

Telegraph House

at Trinity Bay in Newfoundland, where the transatlantic cable reached the North American continent (right).

own cadres of engineers. As British engineers and contractors designed and built a vast number of railway lines, port installations and other works across the globe, their French and American counterparts were also making their mark. It was the Frenchman de Lesseps who constructed the Suez Canal and began the Panama Canal, and French engineers produced two of the most enduring icons of the 19th century – the Eiffel Tower and the Statue of Liberty. Similarly, foreign firms were beginning to challenge British engineering supremacy. Although the 1851 Exhibition displayed Britain as the "work-shop of the world," it also acted as a showcase for leading firms

The Forth Bridge

Built to carry a railway across the Firth of Forth estuary in Scotland, this cantilever bridge was the biggest steel structure in the world when completed in 1890. It took nearly eight years to build and required 54,000 tons of steel; a demonstration of the cantilever principle (top) with upper lattice girders in tension and lower tubular members in compression; the massive cantilevers under construction (above and right).

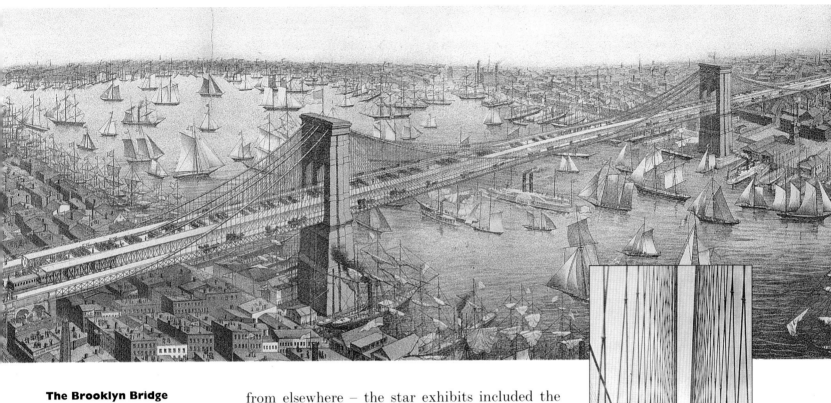

The Brooklyn Bridge

in New York. Another major landmark in bridge construction completed in 1883. With its span of 490 meters, it was half as long again as any bridge built at that time. Its most radical feature was the choice of crucible steel rather than iron for the huge suspension cables, proving the superior tensile strength of steel (above and right).

from elsewhere – the star exhibits included the American McCormick reaper and the huge steel ingot cast at the Krupp works in Essen. Subsequent exhibitions – New York (1853), London (1862), Paris (1867) and Philadelphia (1876) – made it clear that many countries now possessed ingenuity and skills to rival – and frequently surpass – British producers.

The Suez Canal

became a vital waterway not only for trade between Europe and Asia, but also for British imperial interests. The British steamer SS *Malattar* passing through in 1873 (right).

The Great Exhibition of 1851

held in the Crystal Palace in Hyde Park, in London. On the Continent, national exhibitions had been held for many decades but, on the insistence of Prince Albert, the event was transformed into the Exhibition of Works of Industry of All Nations. Over 100,000 objects were shown by 14,000 exhibitors, of whom nearly half were from abroad. One of the great-est attractions was the display of 23 machine tools by Joseph Whitworth & Co (above), one of the world's lead-ing engineering firms. Other exhibits included carpets, tapestries, furniture, household ornaments and even false teeth. The space allocated for American exhibitors in 1851 may not have been filled, but within two years the "American System" was being studied by a committee of which Sir Joseph Whitworth was a member.

The 1862 Exhibition

held in London on the site of what is now the Natural History Museum, the Science Museum and Imperial College. The exhibition buildings were later moved to north London where they formed the first "Alexandra Palace." This photograph (right) of the iron and general hard-ware section shows prize-winning plumbing artifacts in the foreground, together with zinc roof coverings.

Behind is a still by "Henry Pontifex & Sons Coppersmiths and engineers, Manufacturers of apparatus and machinery for brewers and distillers. Also for the East and West India markets." The most noteworthy industrial exhibit was the collection of textile machinery shown by the Platt Brothers.

The 1876 Centennial Exhibition

in Philadelphia, the first major US exposition. All the machinery on show was powered by the huge 1500hp Corliss steam engine (top right). The exhibition was also the setting for the first public demonstration of Alexander Graham Bell's telephone.

The Eiffel Tower

Built to accompany the 1889 Paris Universal Exhibition, Gustave Eiffel's 307 meter tower quickly became one of the most enduring icons of the 19th century, despite being scoffed at by all the cognoscenti and opinion makers beforehand. Both the design and the material used – wrought-iron lattice girders – reflected Eiffel's considerable experience as a bridge builder. Eiffel, in top hat, on the steps of the tower (left).

The Deutsches Museum
in Munich (above): the Kaiser at the opening of the original exhibitions in 1906. Despite the pomp and circumstance, the museum set new standards in demonstrating the interaction between science, technology and industry.

The Patent Office Museum
founded in 1856 (left), housed a large collection of models of noteworthy patents. It also ensured the preservation of such famous relics as Stephenson's *Rocket* and the contents of James Watt's workshop.

Celebrating "Man the Maker"

The work of engineers was celebrated in permanent as well as temporary exhibitions. The tradition began in Paris during the French Revolution, when an ancient monastery was taken over as an educational center for displaying the wonders of engineering. In Britain the South Kensington Museum was established on the back of the Great Exhibition, and the Patent Office Museum was formed to show aspiring inventors the best of what had been done. The science collections of the two museums were amalgamated in 1884 and renamed the Science Museum a year later. There, Stephenson's *Rocket*, Arkwright's spinning frame and other icons of the Industrial Revolution were saved for posterity. After the Centennial Exhibition of 1876 the US Smithsonian Institution displayed American technology in the National Museum Building in the middle of Washington. At the beginning of the 20th century the German Museum for Masterpieces of Science and Technology was set up in Munich, to show the scientific underpinnings of technology.

So elevated was the standing of the engineer, that man came to be defined as such. Discussions of evolution, precipitated by the publication of Charles Darwin's *Origin of Species*, seemed to throw into doubt traditional images of mankind as set apart from all other beings. If man was the result of past selective pressures, the direction of his future evolution was unclear. To some it appeared that the use of machines would be a determining feature of human evolution. In *Creative Evolution*, the extremely popular work by the philosopher Henri Bergson, published in 1907, Homo Sapiens, "Rational Man," became Homo Faber, "Man the Maker."

Entrance doors
of the original South Kensington Museum, from which sprang the Science and the Victoria and Albert Museums. The panels (right) portray British luminaries of science and technology – Watt, Davy and Isaac Newton – alongside great artists of the Italian Renaissance.

*Those who understand the steam engine
and the electric telegraph spend their lives in
trying to replace them with something better.*

G.B.SHAW
Man and Superman, 1903

5 The second industrial revolution

The closing decades of the 19th century saw rapid industrial development throughout parts of Europe and the United States. Some of this was due to the spread of established industries into countries which had seen little change previously. Quite new industries, particularly those associated with the use of chemistry and electricity, also emerged. This period of rapid technological change has been called the "Second Industrial Revolution," a phrase coined by the Scottish town planner and seer Patrick Geddes as early as 1915.

Many technologies whose impact would be felt throughout the 20th century were introduced or invented in the four decades before the start of the First World War – synthetic aspirin, heroin, artificial fibres, radio, the telephone, the electric light bulb, the domestic refrigerator, air conditioning,

A Berliner gramophone of 1890 (above), the first to use a disc rather than a cylinder.

Thermionic valve An experimental bulb made by John Ambrose Fleming in 1889 to investigate the phenomenon of electricity flowing in only one direction. In 1904 he used this first "valve" to detect radio signals. Improved valves were to become the mainstay of electronics in the first half of the 20th century.

98

vacuum cleaners, punchcard machines, gramophones, the gasoline powered car, the airplane and the machine gun. The scientific exploration of new phenomena, particularly by chemists and physicists, the development of measurement skills, chemical analysis and advanced mathematics, the proliferation of

Data processing

Herman Hollerith, who worked on the analysis of the US 1880 census, was appalled by the slowness of the process, so he designed a mechanical tabulator using punched cards to record information. This developed into a system of punchcard devices. A sorter is shown above. His Tabulating Machine Co. later became part of International Business Machines Corporation, IBM. The clerk (left) in an advertisement for Felt and Tarrant comptometers, is using one of the newly available mechanical calculators. Such professionals were known as "computers."

Inventions and inventors

(top left to right) Hiram Maxim, inventor of the machine gun, explaining it to his grandson, 1893; an industrial vacuum cleaner; a woman using the Singer treadle sewing machine in 1902; (middle) a Grahame White biplane at Bisley, 1913; Thomas Edison listening to a cylinder phonograph, 1911; Alexander Graham Bell making a telephone call, 1892; (bottom) Guglielmo Marconi with his wireless, 1896; an 1896 Peugeot driven by Charles Rolls of Rolls-Royce fame; a telegraph operator receiving a message.

professional scientists and the multiplication of specialized instruments had opened up new possibilities. Exploiting these developments required the concerted action of many more people with formal education than previously. Looking back, it is in this period that we see the beginning of research and development departments directly geared towards the generation of new knowledge and the application of scientific research to practical problems.

Chemistry

The chemical industry typified the new approach. Mauve, the first synthetic dye, had been made using an ingredient of coal tar in 1856. The discovery owed something to luck — the chemist had been trying to obtain the malaria drug quinine. The synthesis of alizarin, a synthetic substitute for the red plant dye madder, obtained 13 years later, was achieved by means of much more systematic assaults, carried out almost simultaneously in London and Berlin. Painstaking research identified the chemical structure of the natural coloring ingredient, which was then replicated in the laboratory. Once the potential of modern chemistry became clear, formal scientific knowledge helped drive the pursuit of practical solutions as never before. The age of a vast "R&D" effort had arrived, as exemplified by the epic search for a commercially viable substitute for indigo, a crucial blue dye still familiar today as the popular color of jeans. The task took the German chemical company BASF 20 years and consumed huge resources; at the end of the period it was employing 74 chemists, of whom 20 were involved in "research." The result was that the natural material, supplied by plantations in British India, Indonesia and elsewhere in the tropics, was completely superceded by the synthetic German product.

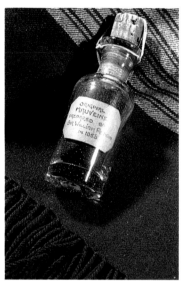

Mauve

One of the first samples of the first synthetic dye discovered by William Perkin in 1856 (left). In the background is a shawl dyed with mauveine, exhibited in 1862.

Home of BASF, Ludwigshafen

The Badische Anilin und SodaFabrik, founded in 1865 as one of Germany's first coal-tar dyestuff factories, went on to produce an enormous range of chemicals (opposite). BASF's indigo laboratory in c.1900 (opposite below) where 17 years' research, beginning in 1880, eventually made possible large-scale synthetic production of the blue dye, indigo (inset). Until then the dye was laboriously extracted from plants as in the Indian factory (right).

German dyes

in enormous variety were being marketed worldwide by 1900. A large collection (left) of samples now in the Science Museum.

"A PICTURE OF HEALTH."

Soap

A product transformed by fierce competition. As consumption rose thanks to more running water and changing habits, firms used advertising and packaging to establish brand identity. Lifebuoy, Plantol and Queen's Honey became household names. Wrapper exchange schemes, introduced by Lever in the 1880s, were also used as a marketing tool.

PLANTOL

SOAP

QUEEN'S HONEY SOAP

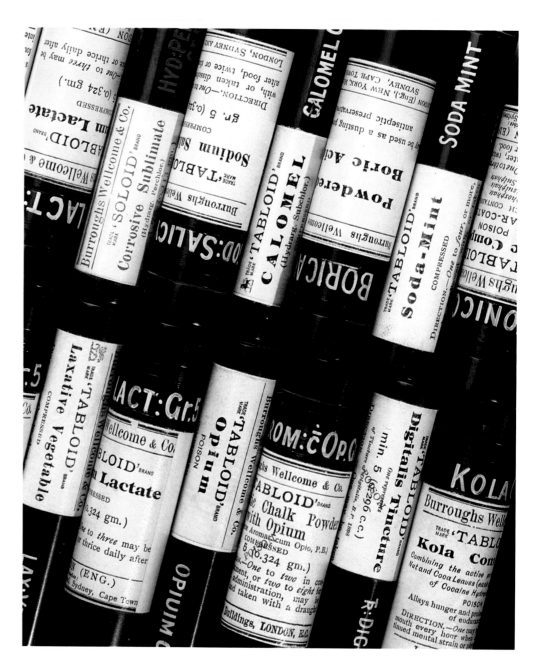

Aspirin

Salicylic acid, extracted from the bark and leaves of willow trees, was noted for its pain-relieving qualities, but side-effects limited its use. In 1897 the German pharmaceutical firm Bayer came up with a similar compound offering the therapeutic benefits without the side-effects. Bayer marketed aspirin in 1899 as a powder, and as a soluble tablet in 1900 (above). It rapidly became the world's bestselling drug.

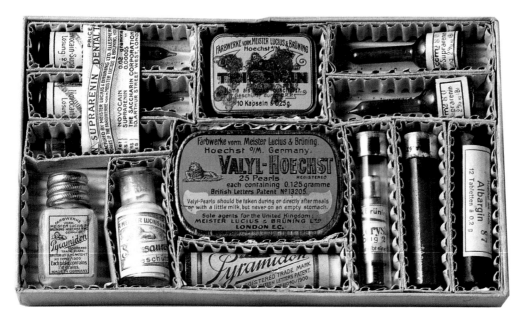

The pharmaceutical industry

As the medical effects of chemicals were better understood, a huge range of new drugs was offered to the public: a chest of ophthalmological drugs, c.1880 (left); a medical supply kit from 1910, made by Burroughs Wellcome & Co. for Scott's Antarctic Expedition (above).

Artificial silk

The first artificial fibre, a form of cellulose nitrate, was patented by the Frenchman Comte de Chardonnet in 1884. This sample (right) was made under his supervision in England in 1896-1900. It was too similar to the explosive guncotton to be safe and was supplanted by viscose, a form of cellulose acetate. In the 1920s such fibres were named "rayon."

Modern chemistry yielded far more than dyes. By the last decades of the 19th century it was possible to speak of a pharmaceutical industry, based mainly in Germany and Switzerland. Early successes included aspirin, Novocaine and barbiturates. Other advances included the development of cellulose products, the first of the artificial fibres. The rapid progress in chemistry had a profound effect on other industries – new standards of engineering were required to construct the complex equipment needed for the commercial exploitation of newly-discovered processes. Early in the 20th century these included the manufacture of ammonia and the refining of petroleum, processes often

Oil

Increasing numbers of cars in the years before World War I saw the rapid expansion of the global oil industry. A view of a Mexican oil rig, 1915 (right); a gusher (above).

involving large quantities of liquids or gases at extreme pressures and temperatures. The chemical industry also provided new materials, such as coatings and lubricants, of crucial importance for all types of engineering. Other products of chemistry included a wider range of artificial fertilizers for agriculture, and new explosives such as dynamite.

Mechanization of farming

gave a huge boost to grain cultivation. The reaping machine proved decisive in opening up the vast potential of the North American prairies. In the US, the McCormick company alone was producing tens of thousands of reapers each year by the 1880s. Major manufacturers, like the Canadian firm Massey-Harris, promoted their products globally, as in this 1890 French advertisement.

Fertilizers

An Italian company advertises its chemical products as bound to increase crop yields, c. 1900 (below).

Electric traction

got its first demonstration with a miniature train installed by Siemens and Halske at a Berlin Trade Fair in 1879 (right). Nearly 100,000 passengers tried out the new experience which was available only on dry days. By 1881 a full-scale system was in use in Berlin.

Hydroelectricity

The hydro-electric power station at Vemork, Norway (left), was the largest in the world when completed in 1911, with ten turbines producing over 100 MW. It supplied the electricity for one of the earliest synthetic ammonia plants, far from any coal mine, the traditional source of energy. Its later use making "heavy water" was commemorated in the film *Heroes of Telemark*.

Electricity from coal

Carville power station near Newcastle, 1907. This was typical of the coal-burning stations in which most electricity was generated. Here, a series of 1500kW Parsons steam turbines was used to drive the dynamos.

Electricity and the internal combustion engine

The growing use of electricity was one of the major features of industrial development during the decades prior to World War I. Scientists had experimented with electricity in the 18th century, and Faraday had constructed an electric motor in 1821. However, the telegraph and electroplating remained its only practical uses until the construction of generators producing substantial current in the 1870s. Thereafter, progress was rapid. Electricity was soon used for lighting and trams, but the real advance came in the 1890s with the possibility of long-distance power transmission. Visionaries, including Thomas Edison in the United States,

Thomas Edison

in old age with the carbon-filament lamp he had invented in 1879.

Light bulb

presented by Edison to the Science Museum in 1880.

An advertisement

in the art nouveau style for light fittings (above) conveys the magic and excitement of electricity transforming the urban evening landscape, with a cluster of bulbs hanging like exotic fruit.

Electric streetcar

Any opening of a new electric street-car line was always a matter of great civic pride. The first streetcar (left), carrying invited guests, arrives at Teddington, West London, in April 1903. Electric streetcars were introduced to London in 1901.

The "Skeleton"

(left) first introduced by Ericsson in 1895, remained in production for 36 years; you turned the handle to signal to the exchange that you wished to make a call.

Early telephones

Alexander Graham Bell obtained a US patent for the telephone in 1876. He demonstrated the new technology to Queen Victoria in 1878 with the model above; the butterstamp-shaped handset was used for both speaking and listening. The dial of the Strowger automatic (below), 1905, allowed people to make local calls through an automatic exchange, without the intervention of a human operator.

Multiple switchboards

enabled local telephone exchanges to service thousands of lines and the network to expand (below). At first, they employed boys, but women were soon found to be more reliable, and another so-called "white blouse" occupation was thus available for them outside such traditional areas as domestic service and factories.

Sebastian de Ferranti in Britain and Walther Rathenau in Germany, built systems for electric power generation and distribution. Alternating power could be transmitted with very low losses. This enabled factory machinery to be run from individual power points, removing the local

The Standard factory

Bishops Green, Coventry, England, 1907 (above). Early cars were produced on a small scale, with all parts fitted by hand, until the advent of Henry Ford's assembly line.

Baby Peugeot

One of a number of small car designs (right), including those of Renault and De Dion-Bouton, which played an important part in the spread of motoring. The *voiturette* cost something less than half a full size car, and could then be run much more economically.

dependency on coal-powered steam engines. Electricity was not the only radical new source of power: engineers were experimenting with different types of engines using gas, oil or petrol. Although gas engines proved impractical for transport, the oil-burning diesel engine proved particularly appropriate for marine use, and began to displace steam power on ships before 1914. Even more important was the petrol-based engine developed by Daimler, which opened the way for the aircraft and the car – the most potent symbols of the 20th century.

Racing

Rapid improvements to the speed and endurance of the car led to the birth of motor sports: (left) the Automobile Club's 1,000 mile trial in 1900; (above) poster for the Indianapolis races, 1909.

1896 Peugeot

(right) the first car owned by Charles Rolls of Rolls-Royce, behind the notorious man with the red flag. A British Act of 1865, which had set a maximum road speed of four miles per hour and required a man with a flag to walk in front, had been designed to thwart the development of the steam bus. As a result of pressure from the small but privileged band of private car owners, the act was soon abolished.

Photography

The announcement in 1839 of the first practicable photographic processes by Daguerre in France and Fox Talbot in England was the precursor of a revolution in visual communication. Most earlier photographic processes in use required photographers to prepare their own sensitive materials. Typically, the freshly prepared solutions were coated onto individual glass plates which were then loaded in a simple plate holder immediately prior to each exposure. In the 1870s gelatine halide dry plates were introduced. The new plates could be produced in

Single-Lens reflex cameras
were only adopted following the introduction of dry plates in the 1870s, and not widely used until the 20th century. They allowed the one lens system to be used for viewing and focusing up to the moment of exposure, and many early examples were designed for use by naturalists. Such cameras were cumbersome, but provided this big-game hunter (above c.1905) with an alternative to the gun.

Snapshots
Early cameras were difficult to use, as the film required careful handling and complex chemical reactions to process the images. However, this changed with Eastman's Kodak camera launched in 1888 (left). A gelatin-coated film on a roll was installed in a compact, inexpensive camera that anyone could use. After the film had been exposed, the camera was sent back to the factory where it was removed, processed and the camera reloaded. The Kodak opened up photography to the general public, and the camera became an important adjunct to leisure (below).

EDISON'S GREATEST MARVEL

THE VITASCOPE

"Wonderful is The Vitascope. Pictures life size and full of color. Makes a thrilling show."
NEWYORK HERALD, April 24, '96.

The movies

In 1895 the Lumière brothers used a "cinématographe" to project a moving film to a large audience. The 1896 No. 8 model is shown on the left. Thomas Edison responded by buying the rights to the competing phantascope, which was hastily marketed as the vitascope (above). Films were made of the new plastic, celluloid, which had many other uses as shown in this advertising artist's whimsy of 1885 (right).

large numbers and stored for long periods. By divorcing the chemistry of photography from picture-taking, a vast new market was opened up. Dry plates were also far more light-sensitive than any earlier process and allowed much shorter exposure times. This meant a tripod was no longer needed, which led to smaller cameras that could be held in the hand. Roll film was widely introduced, allowing rapid consecutive exposures. Roll film also allowed tentative experiments to produce moving pictures to come to fruition, and the establishment of the cinema as a means of mass communication and entertainment. By the beginning of the 20th century the introduction of the half-tone process, permitting shades of gray, led to photographs in books and newspapers.

CELLULOID WATERPROOF COLLARS, CUFFS, & SHIRT BOSOMS.

New sources of innovation

It would be a mistake to see industrial change only in terms of the crashing power of the steam hammer and superior manipulation of the natural world. New ideas concerned with increasing labor efficiency, the new so-called "scientific management," were also developed in manufacturing and in the railway companies.

Vast and costly undertakings, such as transcontinental railways, could hardly be run as family firms so the period witnessed the emergence of the modern corporation in which ownership could be dispersed among numerous shareholders. Although the age is famous for the formidable presence of larger-than-life entrepreneurs like Carnegie, Ford and Edison, the increasing complexity of business activity, nonetheless, led to the separation of control and management, and the rise of boards of directors and professional managers. By concentrating on the most efficient utilization of productive capacity, firms such as the explosives giant Du Pont developed sophisticated cost-accounting techniques which formed the basis for modern management practices. In the growth industries such as oil, rubber, steel and chemicals, large firms were created as a result of mergers and takeovers.

Railway kings
Grand Union Railroad directors meeting in 1868 (left), inside their luxuriously appointed railway carriage.

Plutocrats
The increased scale of business saw the emergence of industrialists and entrepreneurs enjoying huge fortunes: (first row left to right) Friedrich Krupp, German steel and arms manufacturer; Ernst Werner Siemens, head of the electrical engineering giant; John D. Rockefeller, the oil baron; (second row left to right) Walther Rathenau, German industrialist and politician; Frank Woolworth, founder of the chain stores; James Farrell, president of US Steel; (third row left to right) Andrew Carnegie, steel magnate and philanthropist; Thomas Edison; Henry Ford, the father of mass production.

Solvay Conference
of 1911. An international physics conference convened by Belgian chemical magnate, Ernest Solvay. Perhaps the most formidable gathering of scientists ever, the participants included Albert Einstein, who had formulated the Theory of Relativity, Marie Curie, the discoverer of radium, Max Planck, responsible for the Quantum Theory, and Ernest Rutherford, the first scientist to split the atom. Most of those in the picture had won or were to win Nobel prizes.

Migration

New means of transport made possible, and the dream of a better life encouraged over 30 million people to leave Europe between 1880 and 1915. In the USA they created an enormous industrial workforce. A typical poster extolling the attractions of life in the United States, 1919.

The growth of an international economy

With faster communications, the world seemed a much smaller place. The late 19th century witnessed an unprecedented growth in the international economy with cross-border trade, investment and migration. Among the industrialized countries, there was a large degree of mutual interdependence, expressed by the emergence of multinational enterprises. By 1914, firms like Solvay, Siemens and Nobel were operating in several European countries, but the most noticeable trend was the overseas expansion of several dynamic American corporations such as Westinghouse, Standard Oil and International Harvester. The rapid growth of the American corporation created giants which dwarfed their British and European counterparts. Many areas of recent settlement – Australia, New Zealand, South Africa and the open spaces of North America – thrived as suppliers of agricultural and industrial raw materials, and achieved high levels of growth and prosperity.

For most other countries outside Europe, exports to the western world failed to generate wider industrial development. Part of the explanation lies in colonialism, as almost the whole of Africa and large parts of Asia became

Off to Australia

A ship's tender crammed with emigrants takes them out to the steamship which will carry them from Liverpool to Australia, 1913.

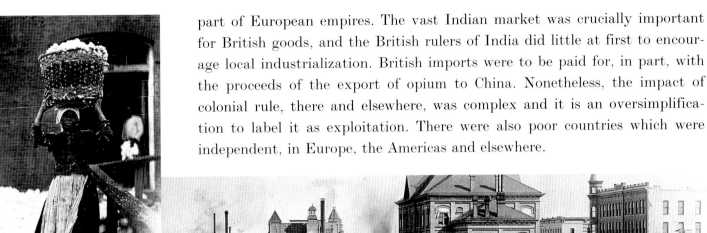

part of European empires. The vast Indian market was crucially important for British goods, and the British rulers of India did little at first to encourage local industrialization. British imports were to be paid for, in part, with the proceeds of the export of opium to China. Nonetheless, the impact of colonial rule, there and elsewhere, was complex and it is an oversimplification to label it as exploitation. There were also poor countries which were independent, in Europe, the Americas and elsewhere.

Raw cotton

exports, from America to Lancashire in particular, soared as demand rose, though much of the work involved had changed little (top). Freight trains transported baled cotton quickly and cheaply from the Deep South to the eastern ports (above) for shipment to Liverpool. Cotton cloth from Lancashire was exported to India, and paid for partly with the profits from opium manufactured there (right) and sold to China.

Unlike the first Industrial Revolution, when Britain had seemed to initiate every major development, the pattern of progress was much more diverse. Britain was still an economic superpower and, as the world's most important source of capital and biggest importer of food, it played a unique role in the international economy. In a growing number of industries, however, technological leadership had passed elsewhere. Firms such as BASF, Bayer and Siemens in Germany dominated the chemical and electrical industries. Several other European countries contained firms that were among the world leaders in their respective fields – Philips, the Dutch electrical giant; Hoffman-La Roche, the Swiss chemical company; Cockerill, the Belgian steelmakers and engineers and Bofors, the Swedish armaments manufacturers. Leading American corporations like Du Pont and General Electric were also highly innovative, with great emphasis placed on research and development. In terms of assets, the largest American firm in 1912 – US Steel – could match the five largest British firms combined.

While many British companies still belonged to the world elite, increasingly stiff competition ate into export markets. Some firms, unable to compete, sought out softer markets. The Oldham firm of Platt Brothers, preeminent in the world market for textile machinery for over half a century, saw record export sales and profits in the years after 1900, but had already been surpassed in size by Draper of Hopetown, Massachusetts. And it was Draper which launched major innovations such as the Northrop automatic loom.

British heavy industry

The locomotive workshops of the North Eastern Railway, Gateshead, near Newcastle, in 1910. These engines, made in one of Britain's oldest industrial areas, were for the company's own use.

Steel in the US

The blast furnaces and rolling mills of the Homestead Steelworks, Pittsburgh, in 1907. Steel production was the measure of industrial might, and Pittsburgh had become a world center (right).

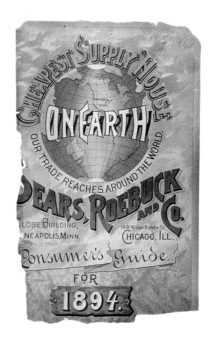

Distribution

The growing scale of production, better communications, and the presence of large concentrated urban markets led to significant changes in distribution and retailing. Goods were sold in new ways. Although traditional small shops survived everywhere, the period saw the emergence of the chain store, with large numbers of identical outlets, typified by Woolworth's in the United States and Lipton's and Sainsbury's in Britain. For the more exclusive end of the market, department stores selling a wide range of products were opening up in major cities. Places like the Bon Marché in Paris, Macy's and Bloomingdale's in New York and Harrods in London became bywords for opulence. Efficient communications and transport enabled the development of sale by mail order, creating retailing giants like Montgomery Ward and Sears, Roebuck. The rapid development of such business underlined

Sears, Roebuck

founded in 1886, sold its goods by mail order catalog (left), offering an alternative to the general store, particularly in the isolated small towns and countryside. The mail order plant, opened in Chicago in 1906, occupied 3 million square feet, and was the world's largest business building.

Harrods

An artist's impression (below) of the famous London department store in 1907. Virtual shopping in 1911 (inset). A well-manned switchboard taking telephone orders at Harrods.

The Shopper

"All the world's a store and all the men and women simply shoppers" – the magazine's motto (above).

Wannamaker's Grand Depot

One of the world's first department stores around 1900 (left and below), it opened in Philadelphia in 1876. It pioneered the use of price tags.

Sainsbury's

Chain stores aimed to have a branch in every main street. Sainsbury's the grocers (above) was founded in 1869 and by 1900 had 48 stores in the London region. Bulk-buying enabled them to sell at lower prices than small shop keepers.

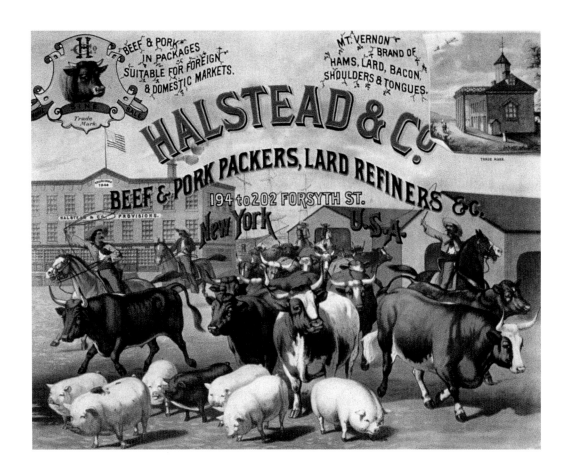

Food from afar
An American beef and pork
packer advertises its products for
consumers at home and abroad.
Refrigeration and ice plants made
long-distance trade in perishables
possible. Ice for packing North
Sea fish is made in Grimsby
(below).

the gradual transformation that industrialization had wrought – if manufacturing was the mainspring of wealth, the consequence was the immense
growth of services which were themselves generating employment and
wealth. Within industrialized countries, technology was now touching the
lives of ordinary people as never before. Manufactured goods became more
affordable, and food was also cheaper as farm machinery, reduced shipping

Bananas

unloaded by American railroad workers in 1910 from refrigerated box cars and probably grown in the new Latin American and Caribbean plantations made viable by them and by refrigerated ships. The precursors of these cars played a big part in sustaining the North in the Civil War.

costs and refrigeration ships allowed the movements of huge quantities of food from the wide open spaces in the Americas and Australasia. With rising living standards in the advanced countries, increasing numbers of industrial workers were able to engage in leisure pursuits. In the workplace, shorter hours and more tolerable conditions made for an easier existence. In the last decades of the 19th century, more women entered the workforce, particularly in the office where typewriters and telephones had revolutionized clerical work. Not everyone experienced such profound changes: in most of the non-industrializing world, and even in most European villages, life still went on much as before.

Take off
Wilbur Wright (left) flies in Glider No. 3 in 1902, the year before his brother Orville made the first powered flight. The Hendon Aviation meeting outside London in 1911 (below). Some spectators got a free view.

Aircraft building

The success of the Frenchman Blériot in crossing the channel in 1909 gave great prominence to the design of his airplane. Here a Blériot-type plane is being constructed in an English factory.

Attitudes towards progress

Advances in science and technology generated contrasting emotions, as people became both fascinated and repelled by the idea of progress. Although mass production ensured uniform quality at prices affordable to unprecedented numbers of people, artists such as William Morris condemned much of the output as ugly. They mourned the disappearance of craftsmanship as much on aesthetic grounds as for the monotony of repetitive labor in factories. Even where acceptance of industrial capitalism was high, as in the United States, hostility could be directed towards the "robber barons" of big business. Other aspects of progress aroused more enthusiasm. Writers such as H. G. Wells, and Jules Verne a little earlier, could thrill readers with stories based on travel in time and space. But if machines could do so much, what guarantee was there that they would not be put to far less benevolent uses? Wells foresaw the use of armored vehicles in war, and Verne described the destructive powers of submarines. Immediately prior to World War I, a new artistic move-

H.G. Wells
appreciated both the benefits of technological change and the destructive possibilities to which it gave rise, predicting in *The War in the Air* (1908) that the application of science to warfare would bring terrible loss of life. Both benefit and threat can be seen in this illustration (left).

H.G. Wells
appreciated both the benefits of technological change and the destructive possibilities to which it gave rise, predicting in *The War in the Air* (1908) that the application of science to warfare would bring terrible loss of life. Both benefit and threat can be seen in this illustration (left).

Futurism
An avant-garde movement originating in Italy in 1909, glorified the vitality of the machine. Their manifesto lauded the "beauty of speed," a theme depicted in Giacomo Balla's *Speed of an Automobile* (right). Composer Luigi Russolu constructed a noise machine for his futurist symphonies (below right).

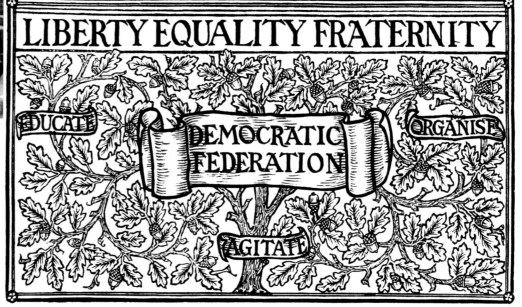

William Morris
designed this membership card, (right) c.1880. He rejected the alienation imposed by the industrial world and called for a return to traditional standards of craftsmanship. A Morris & Co. employee (above) dyes wool by hand in the manner recommended by Morris.

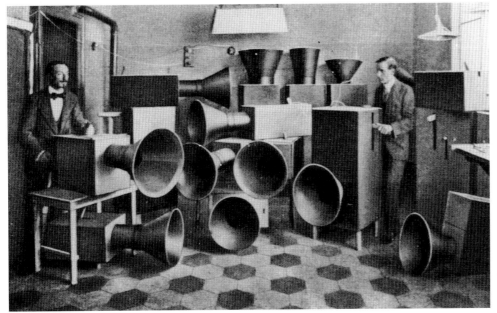

ment – futurism – could extol the speed and vitality of the machine age.

Unlike Britain, where change had been gradual, industrialization in most other European countries seemed far more sudden. Rapid growth in Germany made urbanization there as painful as it had been in Britain, and in Russia industrialization was notorious for the appalling working conditions it brought.

The growing numbers of

industrial workers concentrated in big cities were seen everywhere as an important force for change. Trade unionists sought to improve everyday conditions, while organized socialist political parties advocated the creation of a society based on totally different lines to the bourgeois capitalist mould. Fears of social upheaval brought a mixture of concessions and coercion, as

thinkers, politicians and religious leaders urged the need to address the new evils of the modern world. Astute politicians such as Bismarck brought in welfare schemes to blunt the grievances of workers; other regimes sought to prevent change by increasing repression. Even in laissez-faire Britain, politicians reluctantly admitted

Strikes

The huge increase in the size of the urban industrial working class, coupled with the growth of trades unionism and political mobilization led to a new form of labor unrest and organized strike. The unemployed march in London, 1908 (above). Even in democracies, strikers could be confronted with force. A policeman drawing his baton (left) during the transport workers' strike in Britain in 1912, a year of increasingly bitter labor disputes. Note the load of Argentinian frozen beef in the background.

Match girls

on strike in London in 1888 (right). Lacking the economic and organizational muscle of their male counterparts, women workers found it much more difficult to achieve better working conditions. Poorly paid and badly treated, the match girls were successful thanks to the support they received from political activists.

Dissection class

at the Women's College Hospital, Philadelphia, March 1911. Despite opposition from conservative establishments, women slowly gained access to higher education and entered professions once closed to them. This college was one of the first to open in 1830. By 1900 there were 7,382 female physicians in the United States.

that poverty was still widespread despite a century of unprecedented growth and prosperity, and after 1908 the Liberal government launched several social security measures. Many other countries saw the introduction of factory and antitrust laws and welfare schemes. Just as important was the widespread belief that a modern economy and a modern nation should fund the creation of a trained workforce. The education systems formed then lasted until the enormous transformation in the second half of the 20th century. The growing German university and polytechnic system, the French grandes écoles, the new British technical colleges, night schools and civic universities and the American land grant colleges, graduate schools and liberal arts colleges all testified to the passion for education and training.

Although machinery makes man collectively more lordly in his attitude towards nature, it tends to make the individual man more submissive to his group

BERTRAND RUSSELL
in Whither Mankind, edited by C.A.Beard, 1928

6 The age of the mass

In the first half of the 20th century, people in industrialized countries became particularly conscious of living in a modern age. They associated modernity with science and technology, which they encountered in the mass produced goods they bought, the houses they occupied, and the entertainments they enjoyed. But, for many, technology also seemed to lie behind unwelcome change: was society becoming more machinelike, and was the individual spirit becoming lost in mass society? Would mechanization make work obsolete, and would science perfect warfare and lead to final oblivion? Such fears, aired in Aldous Huxley's *Brave New World*, cohabited uneasily with the belief that science and technology themselves could be the means to realize the promise of mass democracy and overcome the problems of mass society.

Philco Model 444 broadcast receiver, c.1936, known as "People's Sets" (top). Philco was an abbreviation of Philadelphia Storage Battery Company, though the company manufactured worldwide.

The Ford Model T (right). By 1916, they were being turned out at the rate of one every minute.

World War I

Historians talk of a "short 20th century," ending with the fall of the Berlin Wall, but ushered in by the devastating experience of the First World War. This was a "total war" mobilizing civilians at home as well as combatants, and fought with mass-produced armaments. These made it possible to kill and maim on a new scale, whether by machine gun bullets, artillery shells, gas, torpedo attacks or air raids. More than one in eight of the 65 million

A French "75"
served by Americans south of Verdun in September 1918 (below), the first really successful quick-firing gun which had been introduced in 1898. An empty shell case is in mid-air while the next round is fed into the breech. In a static war artillery

became a vital weapon: the British alone used over 170 million shells during the conflict. In this particular battle the Americans used tanks, under the command of Colonel George Patton, and also aircraft.

British soldiers blinded during a German gas attack in 1918. Poison gases such as chlorine, phosgene and mustard gas killed or crippled thousands of troops on both sides of the front (left).

British tanks en route for the front in 1918 (below). The devices they carried were dropped into trenches so that these could then be crossed. Tanks failed to live up to initial expectations, making little impact until the closing stages of the war.

mobilized combatants was killed, and civilians also died in large numbers. The trenches – the most enduring image of the war – were a product of the inability of offensive weapons to counter the defensive capabilities of artillery and the machine gun. Only in 1918 was the stalemate finally broken, when infantry attacks were supported by creeping artillery barrages, tanks and aircraft.

For
EVERY
FIGHTER
a
WOMAN
WORKER
UNITED
WAR
WORK
CAMPAIGN

CARE
for
HER
through The YWCA

Sinews of war
Industry kept up the relentless flow of supplies for the armies. With so many men drafted into the forces, factories recruited large numbers of women. An American poster stressing the importance of the "woman workers" (left). The gun-finishing workshop at Krupp's in Essen (below). A female worker making shells at a Vickers munitions plant (right).

While the infantry endured the trenches, civilians, and especially women, experienced technologies in ways new to them, working in munitions factories and in other areas of work normally monopolized by men, including transport, clerical work and manual labor. An increasingly large share of industrial capacity was given over to the production of armaments and steel. The demands of war accelerated technological change, and some innovations had important civilian applications after 1918, including improvements in road transport, and in air travel, as well as radio communications. The Haber process (1909) for the extraction of nitrogen from the air, used in explosives manufacture during the war, found peacetime application in fertilizer production.

The carnage of war posed both immediate and long-term challenges to medicine, which was itself becoming increasingly technological. Large numbers of wounded were treated in a highly organized system of casualty clearing stations and procedures for "invaliding-out" severely wounded combatants broken in body or spirit. War made substantial demands on nascent medical specialities including orthopaedic surgery, psychiatry and blood transfusion, though wartime innovations did not necessarily find application in peacetime practice. Blood transfusion, for example, did not see widespread civilian use until the Second World War.

An X-ray machine

in a German field surgery tent in 1916. The first use of X-rays in war was in a Graeco-Turkish conflict less than two years after their discovery in 1895.

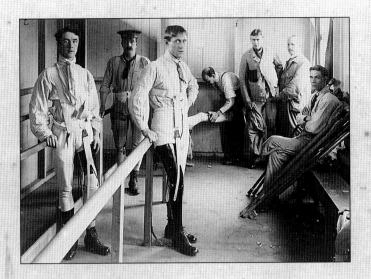

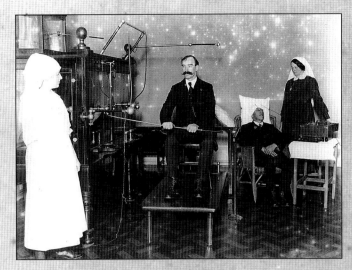

Amputees

at Queen Mary's Hospital in Roehampton, London, an institution set up in 1915 to provide treatment and artificial limbs for the large numbers of wounded servicemen.

Shell shock

British soldiers suffering from Post-Traumatic Stress Disorder undergoing experimental therapy in 1917.

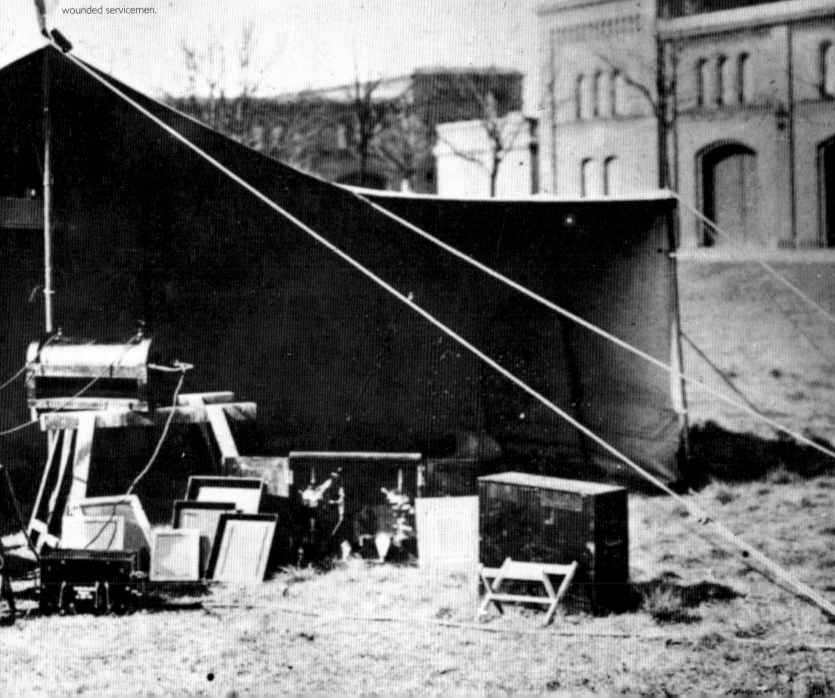

Changing work, mass production, mass consumption

The profound impact of the conflict on the post-war generation was reinforced by other, deeper changes already under way in the world of work. In this "machine age," some made work the subject of art, including films, while others studied it with the intention of rendering production more efficient. Everywhere, contrasts were drawn between traditional industries such as coal mining – portrayed vividly in King Vidor's film *The Citadel* and in the British documentary *Coalface* – and the nature of work in mass-production industry, as satirized in films including Charlie Chaplin's *Modern Times* and René Clair's *A Nous la Liberté*, and in books such as Upton Sinclair's *The Flivver King*.

Increasingly, work itself became the focus of scientific study. Some of the most influential notions about efficiency came from the "scientific management" techniques of Frederick Taylor, who had argued from the 1880s that

The factory as prison:
a still from René Clair's 1931 French comedy *A Nous la Liberté* (top).

Mining
was an industry relatively unchanged by new technology: miners at the coal face of Tilmanstone colliery, Kent in 1930 (above).

production could be speeded up by treating human labor in the same way as materials and energy, the other inputs into industrial production. In his classic work of 1911, *The Principles of Scientific Management*, he presented a system in which time-and-motion study was used to analyze work, and workers were integrated into the factory as efficient components of the "machine." Motion study became a touchstone of increased productivity throughout the interwar period and beyond, both in America and in Europe. It was also deployed to lessen "industrial fatigue," a phenomenon which had come to prominence in the munitions factories of the First World War. In Britain, the National Institute of Industrial Psychology was established with

Time and Motion

a series of photographs showing less (upper row) and more efficient ways (lower row) of operating a hand press, part of a "movement study" carried out in the 1920s by the National Institute of Industrial Psychology in Britain (below).

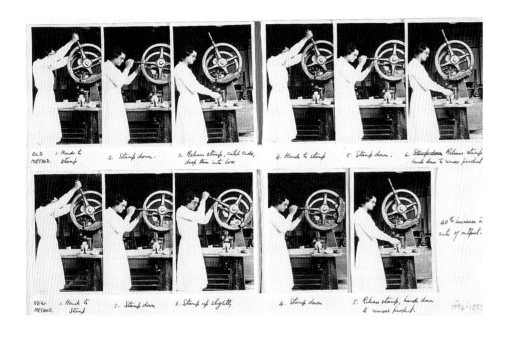

Performance tests

such as the formboard originally devised by American psychologist W.F. Dearborn, attempted to assess general intelligence without employing language (top, right). The Moorrees Form Board, used to select chocolate packers at Rowntree's from the 1920s onwards. The pieces had to be fitted into the right gaps within a given time (bottom, far right). The Timothy Test for mechanical ability, in which a subject would have to correct adjustments made to a small working model (bottom right).

backing from industrialists in 1921. By applying psychological methods, it sought to increase both the productivity and the happiness of workers.

The business most famous for promoting the rationalization of work was the Ford Motor Company, which made the critical moves away from traditional manufacturing to genuine mass production in the decade after 1903. Henry Ford had gathered a talented team of engineers and mechanics who devised new techniques as they sought the most efficient production methods. They took ideas not only from conventional industry, but also adapted methods used in slaughterhouses and grain mills. After years of

Henry Ford

and son Edsel alongside a Model A, the long-awaited successor to the Model T introduced in 1927 (left); the assembly line at the Model T Ford factory at Highland Park in 1913 (below).

The flywheel assembly line

at the Ford factory at Highland Park in 1914 (above).

The disassembly line

A slaughterhouse in Cincinnatti, Ohio in 1873 (below). The methods used in the processing of animal carcasses were said to have influenced the thinking of Ford and his team.

experimentation, a purpose-built factory opened in 1910 at Highland Park near Detroit, with the famous assembly line put into motion three years later. It was designed to produce large quantities of a single standardized product, the Model T car. With assembly lines and conveyors delivering components to the worker at a controlled pace, machine tools designed and positioned specifically to reduce the amount of human movement, careful monitoring of the smooth flow of materials, relentless emphasis on strict adherence to schedules and the utmost accuracy and interchangeability of components, Highland Park itself functioned as an efficient integrated machine, reminiscent of Andrew Ure's vision of an automaton (see p. 36).

(see p. 36)

Large-scale production was scarcely new; after all one of the hallmarks of the Industrial Revolution had been the use of machinery to produce vast quantities of cotton yarn and cloth. Many industrial goods such as sewing machines and bicycles had also been manufactured in large numbers well before 1914. Nevertheless, despite increasing use of interchangeable parts, such production still involved traditional methods of assembly, with skilled craftsmen invariably having to file and finish several components to fit each individual machine. The new form of mass production first became established in the United States partly because there was a sufficiently large and homogeneous market for huge numbers of identical goods.

Ford's Highland Park was able to turn out Model Ts at the rate of one per minute. Between 1909 and 1916, annual output rose from 13,800 to over half a million. Over the same period the sale price of each car fell from $950 to $360. Boosted by intense publicity and sophisticated marketing techniques, Ford's revolution in production methods had made the car available to a far wider market. Equally significant was the openness with which the company sought to share its achievement with other manufacturers. Many elements of Fordism diffused rapidly and, within a few years, assembly line techniques were being used for a wide range of consumer products. By taking mass production to its extreme, however, Ford fell victim to his own success. With 15 million Model Ts sold at an average of over a million per year, the market became saturated. Increasingly, prosperous consumers were less likely to be satisfied by such a uniform product as Henry Ford wished to provide. Rival producers, notably General Motors, understood that high-volume production depended on an equally high volume of consumption, which could no longer be sustained by demand for

The fruits of Fordism

row after row of Model Ts (above). Many car makers had adapted elements of mass production; stacks of car components at the Opel works in Germany, c.1930 (below). Worldwide, much of the market was about to disappear on account of the Depression.

a basic product. Such thinking soon led to the concept of the annual model change, in which consumer interest was maintained by the creation of a sense of novelty and variety. Some of the immediate savings of Fordism were sacrificed in the interests of continuous sales. Thanks to this strategy, General Motors overtook Ford during the late 1920s, and became the largest firm in the world during the following decade. Geared solely to producing the Model

A multi-story car park in Paris in 1930 (right). With the increasing use of cars, special types of building to service and accommodate them began to emerge, including roadside garages and specifically built car parks in major cities.

Boots pharmaceutical factory
in Beeston, Nottingham, a modernist
industrial building completed in 1932 (left),
and designed by Sir Owen Williams.
Reinforced concrete columns bear all the
weight so the external wall can be entirely
of glass and the structure cantilevered out
over the loading bays.

a basic product. Such thinking soon led to the concept of the annual model change, in which consumer interest was maintained by the creation of a sense of novelty and variety. Some of the immediate savings of Fordism were sacrificed in the interests of continuous sales. Thanks to this strategy, General Motors overtook Ford during the late 1920s, and became the largest firm in the world during the following decade. Geared solely to producing the Model

A multi-story car park
in Paris in 1930 (right). With the increasing use of cars, special types of building to service and accommodate them began to emerge, including roadside garages and specifically built car parks in major cities.

Depression, technology and response

This pattern of manufacture and consumption was as yet largely confined to the United States, which enjoyed a decade of unprecedented prosperity while Europe struggled to recover from the ravages of the war. Mass-produced consumer items and entertainments were often identified there as a type of American cultural imperialism. The United States had emerged from the war as the world's sole economic giant, accounting for nearly 50 per cent of global industrial production. This came to an abrupt end in 1929 with the Great Depression, which demonstrated that, despite the impressive advances in manufacturing techniques, economic growth could not be guaranteed indefinitely. By 1933, American industrial output had fallen to half that of 1929, and one in three had no regular work. As the First World War had

Migrant mother

Nipoma, California, a classic image from a farm workers' camp taken by Dorothea Lang in 1936 (left). The Depression was a severe shock for American society, which had grown used to years of rising living standards. Unemployed men line up in a San Francisco employment office, 1938 (below). A soup kitchen for the Bonus Army (bottom), the thousands of impoverished army veterans who descended on Washington in 1932, seeking payment of a long-promised bonus. The authorities took no chances, and cavalry, tanks and tear gas were used to disperse the veterans and their families.

Soviet efforts

to overcome industrial and agricultural backwardness and forge a new society: workers on the assembly line at the Slava tractor works in Stalingrad (below); tractors on a collective in 1930: of great symbolic but little practical value (left). A favorite icon of Socialist Realism, the tractor appeared in all forms of art, including a starring role in a film by Sergei Eisenstein.

Blast furnaces

at Magnitogorsk in the Urals. A showcase project which came to symbolize Stalin's first Five-Year Plan, this metallurgical complex was built from scratch by a combination of volunteer "shock workers" and forced labor (above).

destroyed the mechanisms of the pre-1914 international economy centered around Britain, the American slump exacerbated the problems of other countries. Most became severely indebted to the United States, which was not dependent on imports as Britain had been. For most of the 1930s, recovery was cautious and slow. In the major economies, it was rearmament and ultimately war which allowed a return to full industrial production.

The Depression aggravated technological pessimists' fears for the future. Some believed that mass production and rationalized work practices might produce dangerously long hours of leisure, or alternatively that they had already caused "technological unemployment," putting hundreds of

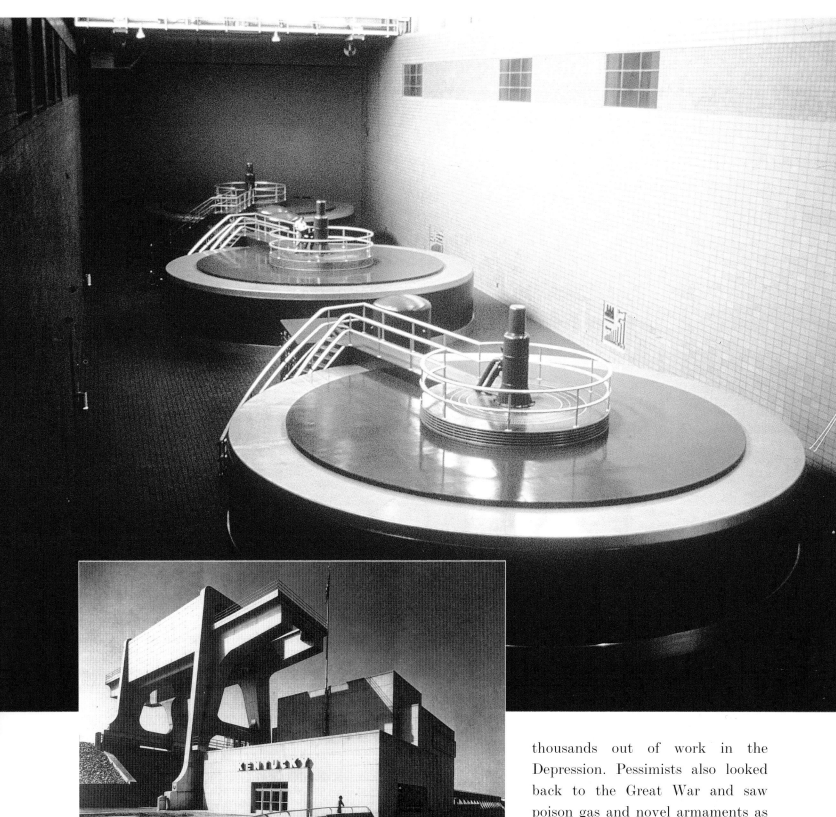

thousands out of work in the Depression. Pessimists also looked back to the Great War and saw poison gas and novel armaments as further signs of a malign science. Some called for a "science holiday," a moratorium on scientific research so that society could catch up with the implications of what had already been discovered. Amongst the poor, millions may have felt that "the machine age" had done nothing to insure a more secure or comfortable life for them.

Tennessee Valley Authority

The generator hall of the powerhouse at the Chickamauga Dam (left); the TVA dam at the northern end of Kentucky Lake (below left).

Ceremonial opening

of the first section of the Reichsautobahn from Munich to Landesgrenze, May 1935 (below). The German army would have preferred the money to be spent on the railways, since it still largely relied on horses.

Liberal governments seemed totally unable to address the economic crisis. Communists could point to the apparent success of the Soviet Union, where a state-sponsored industrialization drive had created an impressive manufacturing base. For the Left, the Depression was a clear sign that the days of capitalism were drawing to a close. In the totalitarian regimes of the Soviet Union, Fascist Italy and Nazi Germany, the glorification of modernity and technological progress were prominent traits. Stalin's grandiose schemes for the transformation of nature or Hitler's autobahns were much lauded examples of what could be achieved by societies supposedly run as finely tuned machines. But Fascist attitudes were not always in favor of technological modernity, as Nazi espousal of wholefood, folk culture and "natural" childbirth clearly demonstrates.

In the United States many engineers felt that only the movement known as "Technocracy" could bring relief from the Depression. For them the precision and efficiency that engineering could bring would replace the waste seen to be endemic in capitalism. Herbert Hoover, US President at the end of the 1920s, was known as the "Great Engineer." His successor Franklin Roosevelt introduced the massive Tennessee Valley Authority scheme, which incorporated over a dozen dams, most of them new, factories and electric power provision. Almost Soviet in the scale of its intervention in the economy, it symbolized for a generation the potential of large scale technological intervention for social betterment within a conventional capitalistic structure.

Mass entertainments

Some have suggested that the revolutionary potential inherent in the distress caused by the Depression was neutralized by the new distractions available for ordinary men and women. The public which made up the market for mass-produced goods also constituted the audience for new mass media. In the printed word, sound recording, radio and cinema, the interwar period saw an efflorescence of popular entertainment. Each of these media was reliant on a series of technical developments which built on the innovations of the previous century, just as their contents owed much to the popular forms of the past. In newspaper publishing, the Linotype machine, which multiplied the productivity of typesetters twelve-fold when introduced in the 1880s, permitted proprietors to reach the greatly enlarged literate public. The American "yellow press" of the 1890s was in the vanguard of a new generation of popular newspapers which, increasingly, introduced new visual elements: banner headlines, cartoons, display advertisements and photographs. In the 1920s many of these changes were aided by innovations in printing machinery and new inks, just as the introduction of lightweight cameras, especially the Leica, along with the flash bulb and the telephoto lens, revolutionized press photography. Whole new publication genres, such as the photojournalistic magazines *Picture Post* and *Life*, were created to exploit these new techniques.

This period has been called "the age of the dream palace," referring to

Photojournalism

The first issue of *Picture Post* from October 1938 (below).

Advertising poster

for Emil Gerasch, art printers of Leipzig, c.1925 (above), by Ludwig Hohlwein, the leading German commercial artist of the 1920s.

The dream palace

The cinema became one of the most popular forms of communal entertainment. Studios made full use of marketing techniques to arouse interest in their latest productions: a poster for the 1931 blockbuster *Frankenstein* (above). Cinemas sought to create a unique experience with distinctive architecture and plush interiors: floodlit and outlined in neon, a cinema announces the arrival of Charlie Chaplin's *Modern Times* (above right). Films had particular appeal for the young: schoolboys watching *Things to Come* in 1936 (right). Each of these films was commenting on the place of technology.

Printing press

Large, high-volume photogravure printing presses, like this German one (left), began to be used for newspaper and magazine printing just before the First World War. By the 1920s and 30s they were much in evidence, producing mass-circulation illustrated magazines.

the cinema building craze that served an annual audience figure of 987 million by 1939 in Britain alone. By the beginning of the First World War, Hollywood already dominated cinema production so that in 1926 fewer than 5 per cent of films shown in Britain were produced locally. Even in rural districts, travelling projectionists, taking advantage of affordable motor transport and new smaller gauges of film and more lightweight projectors, could bring the latest *Frankenstein* to amaze communities of people whose own direct experience might often extend little beyond their immediate region.

The gramophone became increasingly popular; in Britain sales rose from 4 to 26 per cent of musical instrument sales between 1907 and 1926, and *The Times* was able to report in 1929 that "in every street, in every block of flats it is usual now to hear half-a-dozen or more gramophones all making different noises at once." The radio acted as a powerful source of entertainment, instruction and even indoctrination. By 1930, the crystal sets on which most early enthusiasts listened had been superseded by valve sets with greater fidelity, and which no longer limited listeners to using headphones.

The gramophone

Listening to recorded music became a favorite leisure pastime, but also a major source of irritation. The HMV gramophone from 1923 featured a Lumière pleated diaphragm, a short-lived experimental acoustic device used instead of a horn to produce an amplified sound (left).

NO NEEDLESS NOISE

JOIN THE ANTI-NOISE LEAGUE

The Anti-Noise League

A 1935 advertisement for this British pressure group campaigning against a prime cause of stress in the "machine age."

BBC Television

began broadcasting in 1936. In September 1937, a football match involving Arsenal, the leading English side of the decade, was the first to be televised. Here the players are photographed examining one of the cameras.

The machine-age city

This first age of mass production and consumption reached its apogee in the great cities which increasingly became fringed with large suburbs growing octopus-like along radial roads, served by streetcars, railways or bus services. Rising car ownership led to changes in the layout of cities and suburbs. Technological optimists and pessimists turned to cities to support their beliefs. In the interwar imagination, towns might be gleaming technological utopias, as seen in the film *Things to Come*. Alternatively, they were seen as the seats of technological unemployment, as "Worktown," the thinly disguised version of Bolton in the northwest of Britain, explored by the amateur anthropological organization Mass Observation. Modernist architects, preeminently Le Corbusier, as well as designers of the German Bauhaus school, imagined a new way of living in rational surroundings; for them, a city, as much as a house, was a "machine for living in." Suppliers of gas and electricity promoted their products as rational and modern alternatives to coal, still the dominant fuel. New appliances, including washing machines, irons, vacuum cleaners, water heaters and cookers, changed patterns of housework and leisure.

A streamlined Hudson locomotive

on the New York Central Railway (left), designed by Henry Dreyfuss, the pioneering industrial designer. Dreyfuss helped transform the design of many everyday products, from Bell telephones to Polaroid cameras and Deere & Co. tractors.

Quarry Hill Flats in Leeds

a local authority steel-framed project modelled on successful examples of worker housing from Vienna. Built to house 3,000 people, the flats were the largest of their type in Europe when completed in 1938 (right).

The machine-age city

This first age of mass production and consumption reached its apogee in the great cities which increasingly became fringed with large suburbs growing octopus-like along radial roads, served by streetcars, railways or bus services. Rising car ownership led to changes in the layout of cities and suburbs. Technological optimists and pessimists turned to cities to support their beliefs. In the interwar imagination, towns might be gleaming technological utopias, as seen in the film *Things to Come*. Alternatively, they were seen as the seats of technological unemployment, as "Worktown," the thinly disguised version of Bolton in the northwest of Britain, explored by the amateur anthropological organization Mass Observation. Modernist architects, preeminently Le Corbusier, as well as designers of the German Bauhaus school, imagined a new way of living in rational surroundings; for them, a city, as much as a house, was a "machine for living in." Suppliers of gas and electricity promoted their products as rational and modern alternatives to coal, still the dominant fuel. New appliances, including washing machines, irons, vacuum cleaners, water heaters and cookers, changed patterns of housework and leisure.

A streamlined Hudson locomotive

on the New York Central Railway (left), designed by Henry Dreyfuss, the pioneering industrial designer. Dreyfuss helped transform the design of many everyday products, from Bell telephones to Polaroid cameras and Deere & Co. tractors.

Quarry Hill Flats in Leeds

a local authority steel-framed project modelled on successful examples of worker housing from Vienna. Built to house 3,000 people, the flats were the largest of their type in Europe when completed in 1938 (right).

Everytown, the city of 2036

from Alexander Korda's *Things to Come* (1936). Based on the novel by H.G. Wells, the film portrayed an enlightened technocratic future in which society had learned to overcome the traumas of the 20th century, and to make life "lovelier and lovelier" (left).

Futuristic design

A telephone booth using sound-proof glass, Warsaw 1938 (below).

The Illinois Building

at the Chicago World's Fair of 1933, a celebration of "A Century of Progress" (right). The exhibition attracted nearly 40 million visitors.

Ray therapy

in 1930 (below). Ultraviolet therapy was commonly used as a treatment for rickets and tuberculosis, and was regarded as particularly beneficial for poor children who lived in polluted cities, making up for a lack of sunlight and so of Vitamin D. Many local authorities in impoverished city boroughs provided light clinics for their inhabitants.

Technology fired the imagination with unique feats such as Lindbergh's solo flight across the Atlantic. It also promised to revolutionize the everyday lives of individuals, whether via mass production, new materials such as synthetic plastics and stainless steel, or anticipated applications for new discoveries in science. But not everyone was impressed; in Britain the King's Physician, Lord Horder, commented acidicly that "Bound up with, and inseparable from, this hustle that has caught us all, and the machinery to which we have become the slaves, is the collecting of ourselves into masses which we call cities."

> *No nation which expects to be the leader of other nations can expect to stay behind in the race for space*

PRESIDENT JOHN F. KENNEDY
Rice University, 12 September 1962

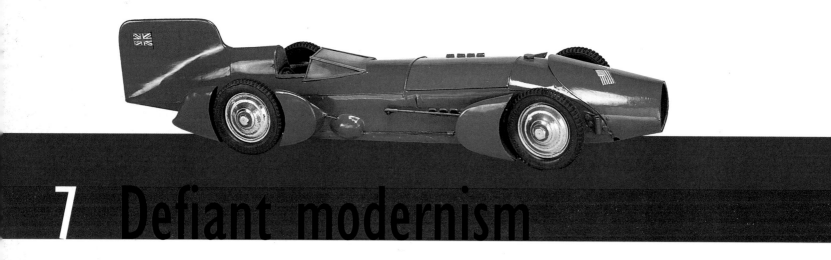

7 Defiant modernism

Governments and the media have periodically exploited achievements in science and technology as powerful symbols of global status. This was particularly evident from the late 1930s to the late 1960s. Even before World War II, in an era of heightening international tensions, the great world fairs and rivalry for speed records on land, water and in the air had linked national pride with technological achievement. The 1939–40 World's Fair in New York and its counterpart held at the same time in San Francisco provided dramatic showcases for nations and private companies to display their claims to technological achievement. During the war, both sides

World speed record breakers from the decade of streamlining: a model of Malcolm Campbell's *Bluebird* car (246.09 mph, 1931) (above) and the *Mallard* steam locomotive (126 mph, 1938) designed by Sir Nigel Gresley for the London and North Eastern Railway (right).

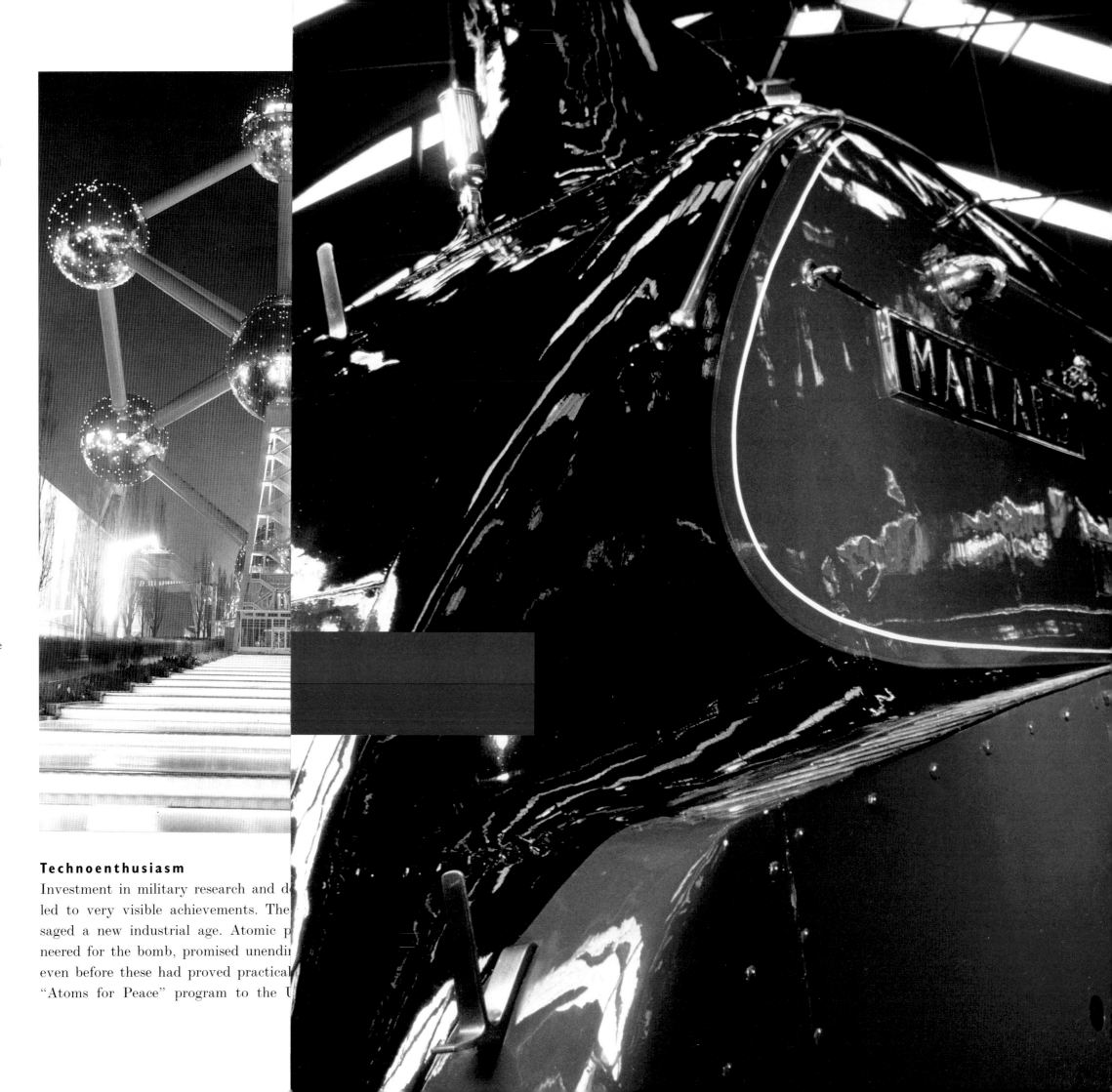

The Perisphere
a 58.5 meter hollow ball housing the New York World's Fair "Theme Center" (left). It was subsequently demolished. The fair, celebrating "the world of tomorrow," opened in April 1939 and ran until October 1940. Germany and China were the only two major countries not officially represented there.

The Atomium
The peaceful uses of atomic power were central to the 1958 World's Fair in Brussels. They were symbolized by this towering sculpture based on the crystalline structure of iron magnified 165 billion times linearly (right). It is 102 meters high and composed of 9 steel balls, representing atoms, each with a diameter of 18 meters. The exhibition provided an opportunity for the Soviet Union to show off its scientific and technological prowess. "Beeps" from the six-month-old Sputnik were broadcast.

Technoenthusiasm

Investment in military research and d[...]
led to very visible achievements. The [...]
saged a new industrial age. Atomic p[...]
neered for the bomb, promised unendi[...]
even before these had proved practica[...]
"Atoms for Peace" program to the U[...]

devoted great efforts to leapfrogging each other's weapons' cap[...] just for military purposes but also to instill fear and respe[...] numerous developments, the most awesome was the atomic [...] neered in the USA, through the Manhattan Project. During the [...] governments, led by the United States, made substantial inve[...] research and development. The proportion of the US gross nat[...] uct devoted to such government support increased fivefold be[...] and 1950, and another fivefold by 1963. The space race, from [...] of Sputnik by the Soviet Union in 1957 to the American M[...] twelve years later, most vividly illustrated international [...] through technology.

N° 242

SERIAL N° 239

TENT APPLIED FOR

Computers

Instead of today's "hard disks," the information on early computers was stored on paper tape or magnetic tape. In the cinema and in the popular imagination the large mechanical tape drives became symbols of these electronic brains. The 1963 Elliot was an early transistorized computer which used 35 mm film stock coated with magnetic material for information storage (overall and left). The valve-based Ferranti Pegasus (far left), made between 1956 and 1962, which required special refrigerated cooling and weighed 2000 lb, used paper tape for programs and data and had a magnetic drum with a memory of 33 kilobits. A modern personal computer might typically have magnetic storage 10 million times greater.

Spitfires and Hurricanes

symbols of pride in British technology (right). The
Hurricane (above) represented a sophisticated
evolution of traditional aircraft construction technique.
Metal tubes and frames covered in plastic soaked
("doped") fabric formed an immensely strong composite
structure. By contrast the Spitfire (below) used the new
"stressed skin" construction, developed for the lastest
US airliners, in which a continuous outer light alloy
surface shared a large proportion of the flight loads. This
led to a lighter aircraft with higher performance.

Radar and fighter

A radar station receiving room (left) feeding
information to fighter control, which then
dispatched Spitfires like the one flying beneath a
Heinkel's rear gun turret (right) in 1941.

British technopride

The British took particular pride in their wartime developments such as radar, the jet engine, and penicillin. Radar had been developed before the war in several countries. In Britain there was a distinctive development to answer a growing concern about vulnerability to air raids from the Continent – integrating the evidence from each separate radar station to form a national picture, and combining this with centralized air raid and fighter control embracing all aerodromes and ultimately the defending fighter pilots. This "Chain Home" system used arrays of large static aerials and, though not outstanding in its electronic design, it created for the first time a sophisticated reporting and control system.

Once enemy raids could be detected by radar, fighter aircraft – Spitfires and Hurricanes designed by R. J. Mitchell and Sydney Camm, respectively – could be "scrambled" to intercept German bombers. In the early 1940s, the development of the cavity magnetron at Birmingham University made possible more compact high-frequency radar systems to be mounted on aircraft. These proved of particular importance in the search for U-boats attacking convoys importing food and essential supplies from America – the Battle of the Atlantic. The original

prototype of the magnetron was taken to the United States for further development as part of an agreement to share technology. At the end of the war the success of the jet (though too late to be of military significance in the struggle) promised even faster and more potent British aircraft.

There was a tendency to personalize technological achievement and make heroes of inventors. The tenacity of Frank Whittle, inventor of the jet engine, who persisted in the face of "official indifference," became legendary and formed part of the new British self-image. The "discovery" of penicillin too was personalized and special status was given, in British accounts, to the initial identification of the penicillin mould by Alexander Fleming in 1928, although it was not until 1940–41 that a team at Oxford University of more than a dozen people, including Ernst Chain and Howard Florey, managed to purify the drug and to evaluate it as a medicine. And it was only after the techniques of industrial production were established, largely by a major American research effort in industry and government laboratories, that the new drug could be made widely available to soldiers in 1944, and shortly afterwards to civilians.

Great Britons

The German-born Ernst Chain (top left), whose work with Florey led to the separation of penicillin and identification of its curative properties; John Cockroft (top center) leader of Britain's post-war nuclear research program and one of the team to build the world's first accelerator used in nuclear physics, 1932; Reginald Mitchell, designer of the Spitfire (top right); Barnes Wallis, the inventor of the bouncing bomb used to destroy German dams during World War II (middle left); Frank Whittle, inventor of the jet engine (middle center); H.E. Bishop, designer of the de Havilland Comet (middle right); the Australian-born Howard Florey (bottom left); Alexander Fleming, who discovered penicillin and won the 1945 Nobel Prize for medicine with Florey and Chain (bottom center); Christopher Cockerell, inventor of the hovercraft (bottom right).

International stars

Carl Djerassi, the Austrian-born American who first synthesized the oral contraceptive – the "pill" (top left); Wernher von Braun, the German rocket scientist responsible for the V2 and, ultimately, the rockets used in the American space program (bottom left); Andrei Tupolev, the Soviet aircraft designer responsible for many innovative military and civilian aircraft (top right); Jonas Salk, the American microbiologist who discovered a vaccine for the viral disease polio (bottom right).

After 1945 technology came to provide a way of thinking about the world, not just for a few intellectuals, but for millions trying to make sense of the new age. European countries sought to define their identity under the shadow of US dominance in politics, industry and culture. American factories were the largest and most modern. Americans were better fed and widely perceived to have easy access to consumer goods which were scarce in early postwar Europe. Imperial ideas, which had dominated much British thinking since the 1880s, began to fade, while colonies in Asia and Africa gained independence. Celebrating technological achievement provided an alternative way of expressing pride in the past and a hope for the future. American dependence on British technological initiatives was highlighted. These included television (Britain launched the world's first regular

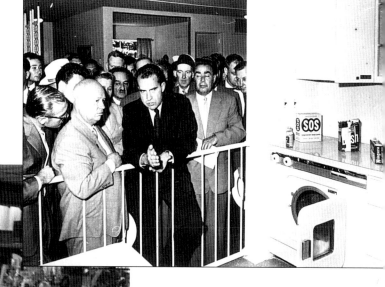

The Great Kitchen Debate

Vice President Richard Nixon and Nikita Khrushchev discuss the merits of their respective systems at the first ever US exhibit in Moscow in 1960. Each boasted how their people would have access to such household goods (above). An American airman inspects washing machines at the Air Force PX (Post Exchange) in South Ruislip, Middlesex in 1955 (left). The picture illustrated a bitter *Picture Post* article comparing the much less luxurious products available to British servicemen.

American innovations

A drive-in bank in Brussels in 1957, one of Europe's first examples of a technical convenience familiar to American towns (above). A British milk bar in 1960, a direct derivative of the US drugstore soda fountain, aimed at the teenage consumer – another phenomenon first seen in America. This picture (above, right) illustrated a *Picture Post* article entitled "Sex and the Citizen." Cathy Norris (right) cooks hamburgers on an electric stove powered by atomic energy at West Milton, New York, celebrating the first commercial use of nuclear power, provided by a nuclear reactor on board the navy submarine *Sea Wolf*, August 10, 1955.

broadcasting system in 1936), the new plastic polyethylene, and penicillin. A National Research Development Corporation was established in 1948 to take advantage of the results of government-funded research.

The British aircraft industry, which had played a crucial role in the Second World War, had a special importance. The earliest civilian jet air-liner, the British Comet, first flew in 1949, and through the 1950s governments invested large sums in developing world class military aircraft, including such successes as the English Electric Canberra, the Lightning and the Hawker Hunter. In 1947 numbers in the industry had been down to 167,000 but by 1955 they were up to nearly 300,000. Vickers, Avro and Handley Page all built nuclear bombers. When partnership with the US in

Vulcan bombers

developed to drop Britain's nuclear weapons (above). First delivered in 1956, they were one of three marques of "V-Bomber" – along with the Victor and Valiant – designed to fly to the Soviet Union and back, at a speed approaching that of sound. Because the performance required was at the limits of understanding, three types were developed, to insure that at least one would be successful.

nuclear weapons was abruptly terminated in 1945, the British moved ahead on their own and in 1953 became the world's third nuclear power. Britain's independently developed atomic bomb was presented both as a deterrent to future wars and a means of avoiding dominance by the USA. Calder Hall, in northern England, opened in 1956, was the focus of national pride as the world's first large-scale civil nuclear power station, although another primary purpose was to produce plutonium intended for atomic weapons. Substantial investment went into building a nuclear industry using gas cooled reactors, although this type failed to win more than a few export orders. The world nuclear power market came to be dominated instead by the American designed light-water reactors, particularly pressurized water reactors, based originally on those in submarines.

The rapidly developing technologies of medical care became available to many in Western countries as each instituted its own version of the Welfare

Atomic destruction

A series of frames (below) showing the total obliteration of a house near the blast site of a nuclear test carried out in Nevada in 1952.

Atomic energy

The opening of Calder Hall in northern England (right), the pioneering nuclear power station, by the Queen in October 1956. Spent fuel was processed and plutonium extracted on a neighboring site at a plant then known as Windscale. Today the sites are known collectively as Sellafield.

State. In Britain the implementation of the 1942 Beveridge Report into post-war social security included the introduction of the National Health Service in 1948, providing free medical treatment. To some in Britain, the aspirations to build a New Jerusalem, with substantial spending on social projects, were excessive and to the detriment of industrial and economic development. Many, however, did not accept that the two should be mutually exclusive. For them the social programs were inseparably welded to the overall technocratic aspirations of the State, and the experience of centralized direction during the war suggested that such programs were achievable. It was argued that if the UK could have a supersonic 1000 mph fighter, then it could certainly also arrange mass X-ray screening and decent housing. Children wearing effective if unfashionable NHS glasses became symbols of the new tomorrow.

A prime opportunity to celebrate the new technology was provided by the Festival of Britain, which opened in 1951. This morale boosting exhibition had three main sites: the display focusing on the new Britain on the South Bank, where the main building was entitled "The Dome of Discovery," another on heavy engineering in Glasgow and a third on science in a newly built block at the west end of the Science Museum. The

The National Health Service

Mass X-ray screening in Croydon, part of the struggle against tuberculosis. Early detection and treatment reduced the risk not only of mortality, but also of the patient becoming infectious (above). A Glasgow boy wearing NHS glasses, 1951 (left).

Council housing

paid for by the state (below), was one of the main ingredients of the post-war welfare programs, tackling chronic overcrowding, replacing slums and the half-million houses destroyed in the war. Moss Heights in Glasgow, a project completed in 1953.

Infrastructure

During the 1950s, Western countries greatly expanded their networks of high-speed roads. In Britain, the first long distance motorway, the M1, was opened in 1959 (right). A television set bought to watch the coronation of Elizabeth II in 1953, a major stimulus to the broadcasting network and TV ownership in Britain (below). Ziggurat – stepped – student residences, designed by Sir Denys Lasdun at the University of East Anglia, one of the first of Britain's new universities of the 1960s (bottom). The building of such universities and the explosion in opportunities for higher education characterized industrialized countries at the time.

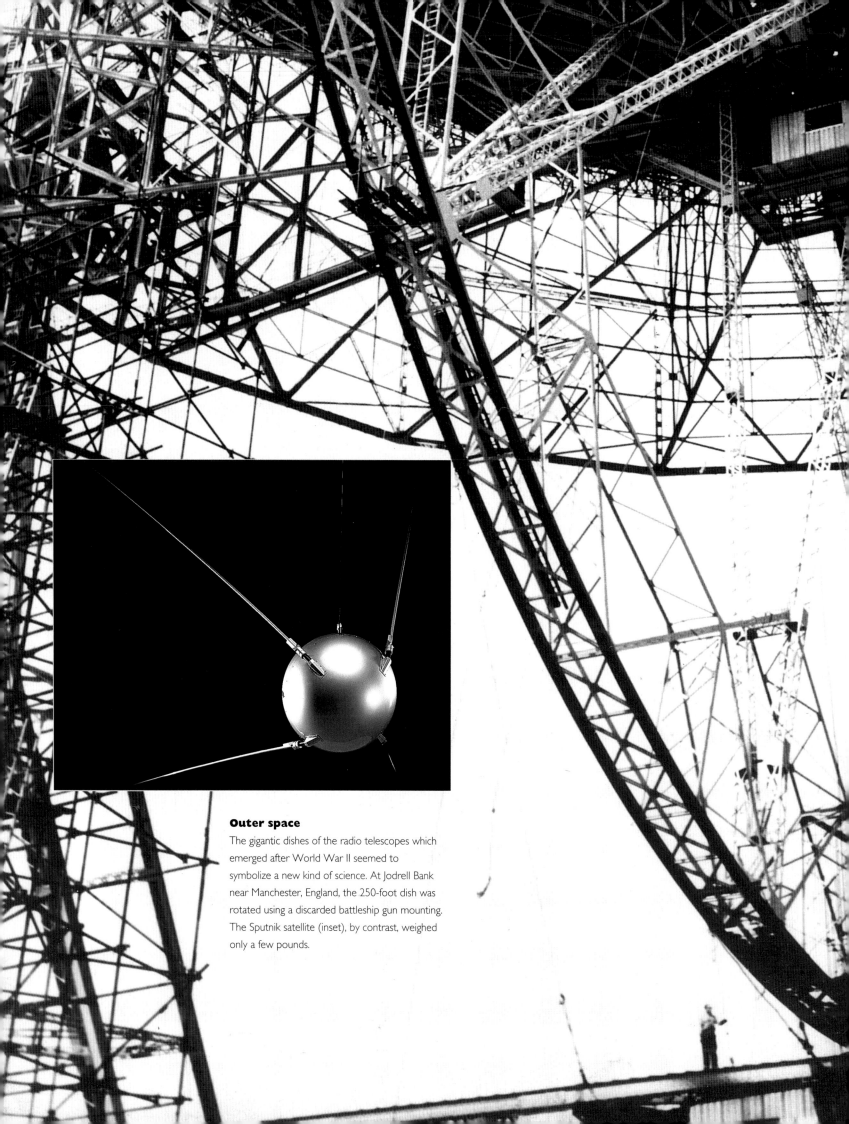

Outer space

The gigantic dishes of the radio telescopes which emerged after World War II seemed to symbolize a new kind of science. At Jodrell Bank near Manchester, England, the 250-foot dish was rotated using a discarded battleship gun mounting. The Sputnik satellite (inset), by contrast, weighed only a few pounds.

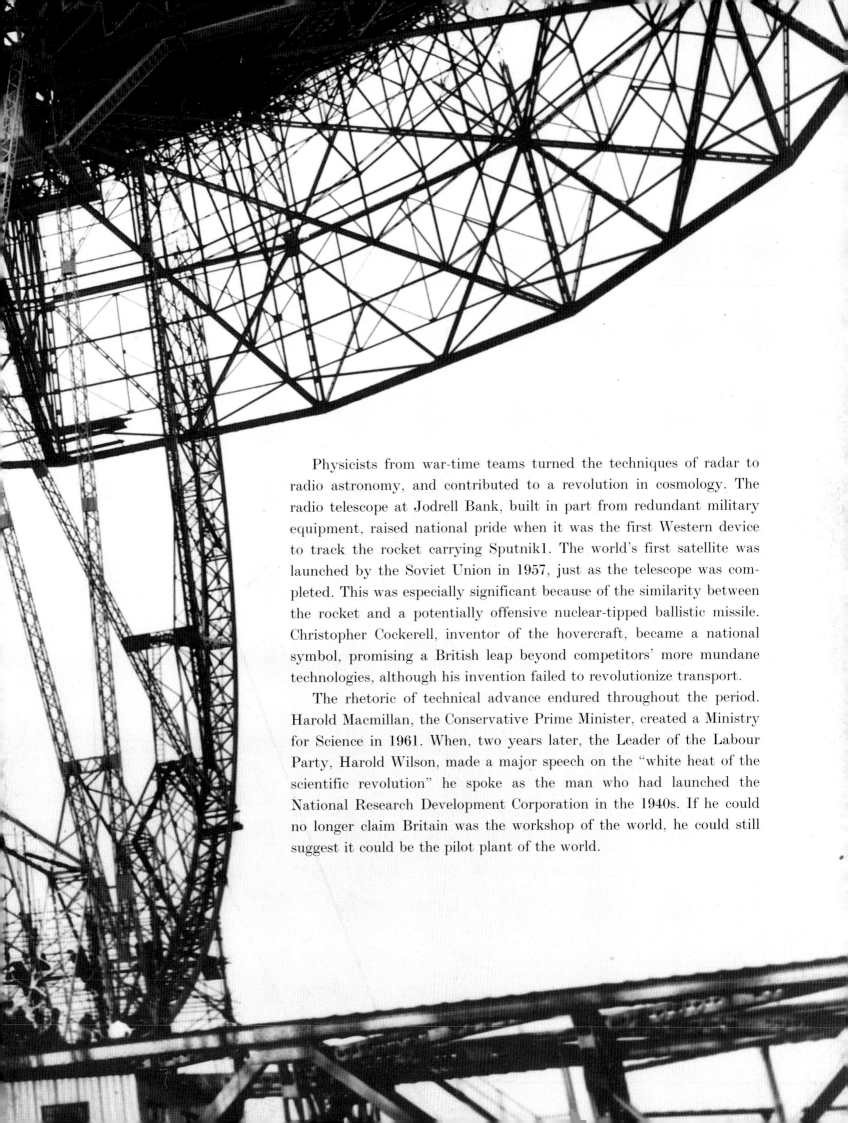

Physicists from war-time teams turned the techniques of radar to radio astronomy, and contributed to a revolution in cosmology. The radio telescope at Jodrell Bank, built in part from redundant military equipment, raised national pride when it was the first Western device to track the rocket carrying Sputnik1. The world's first satellite was launched by the Soviet Union in 1957, just as the telescope was completed. This was especially significant because of the similarity between the rocket and a potentially offensive nuclear-tipped ballistic missile. Christopher Cockerell, inventor of the hovercraft, became a national symbol, promising a British leap beyond competitors' more mundane technologies, although his invention failed to revolutionize transport.

The rhetoric of technical advance endured throughout the period. Harold Macmillan, the Conservative Prime Minister, created a Ministry for Science in 1961. When, two years later, the Leader of the Labour Party, Harold Wilson, made a major speech on the "white heat of the scientific revolution" he spoke as the man who had launched the National Research Development Corporation in the 1940s. If he could no longer claim Britain was the workshop of the world, he could still suggest it could be the pilot plant of the world.

Disillusion and 1968

During the 1950s and 60s the lead in many technologies in which the British played a pioneering
role seemed to slip away. The crashes of the Comet
jet airliners in the early 1950s meant that it would
be the American Boeing 707 which transformed
tourism and business travel. In 1960 the Blue
Streak rocket, the intended delivery vehicle for
Britain's independent nuclear deterrent, was cancelled and replaced by America's Polaris. Five
years later, the TSR2 fighter, Britain's last independent large strike aircraft project, was cancelled
shortly after the first test flight of the prototype.

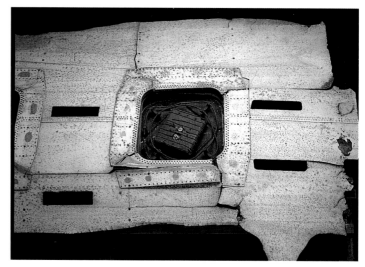

Five thousand were made redundant at the British Aircraft Corporation, the
consortium formed to build it, but at least its engine was able to be developed to power the Anglo-French Concorde. This supersonic airliner, a sensu-
ously beautiful product of Cold War technology, almost went the same way,
before making its first flight in 1969. It was the Boeing 747 jumbo, which

flew for the first time the same year, that made big profits for manufacturers and airlines. Increasingly, technological development for its own sake was questioned by governments concerned to maintain growth rates. Solving the problems of disseminating new technology took priority over the glamour of sponsoring world beating invention. The British, through a new Ministry of Technology, attempted to emulate the success of the US Advanced Research Products Agency founded in 1958. One exception to this trend was the Hawker Siddeley Harrier jump jet developed in the 1960s by the firm on its own initiative because it was running out of work.

Aviation

A model of the British TSR2 used in wind tunnel tests (above). Wreckage from a de Havilland Comet which crashed into the Mediterranean in January 1954 (below left), one of a series of such Comet crashes caused by metal fatigue. Cracks started in the corners of windows or of the radio compass hatch shown here. (Left) a dress rehearsal on board a Boeing 747, prior to the launching of the New York to London service in 1970.

The CT scanner

The Computed Tomography brain scanner from the Atkinson Morley Hospital in Wimbledon, the first to be installed for trials on patients in 1971. Godfrey Hounsfield had used his background in radar to develop scanning, which produced a three-dimensional image of internal organs for the first time. By 1977, there were 1,130 machines in operation.

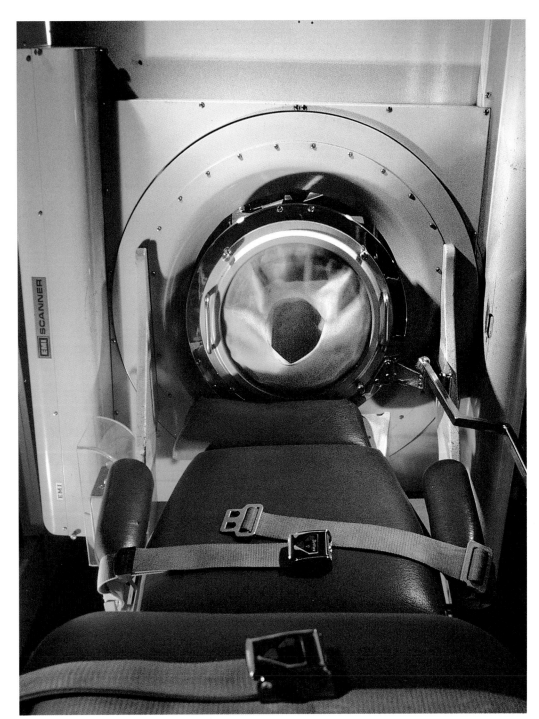

Tiger economies

Grand corporate edifices testified to the confidence felt throughout the Pacific Rim. The Hong Kong and Shanghai Bank Building in Hong Kong (left), designed by Sir Norman Foster and completed in 1986, was claimed to have cost more to erect than any other building. Moving mirrors tracking the sun angle light into its central well. Its ground floor level has a public open space from which some of the longest freestanding escalators in the world take customers to the floors above. I. M. Pei, the architect responsible for the Pyramide du Louvre, designed the 72 story Bank of China (above), with triangular bracing to withstand the typhoons of Hong Kong.

Formica
Cleaning a kitchen was made much easier with such new plastic laminated surfaces. Transparent and water-resistant melamine formaldehyde polymer, available from 1938, was used to impregnate and protect patterned surfaces. Almost immediately it became a symbol of modernity in drugstores and lunch counters.

Tupperware
Single-impression mould used in the injection moulding of molten polyetheylene to make a mix-n-stor container, 1965 (right). Polyethylene was invented in the 1930s but only became available for civilian use after 1945. Its flexibility contrasted with the rigidity of most pre-war plastics. The American Earl Tupper invented the Tupperware Party to help women learn how to apply the tightly fitting lids in a social setting, but also to act as an ambiance conducive to sales of the product.

The Petronas Twin Towers
(1483 ft/452 m) in Kuala Lumpur, Malaysia (below) were designed by Cesar Pelli in 1996, and are still the world's tallest buildings. Previously, the Sears Tower, Chicago, had held the title for 23 years.

"Mac-Country: 100% lean chopped pork with barbecue sauce"
The collapse of the Soviet bloc created two-speed economies in which many enjoyed new opportunities while millions faced severe hardship and an uncertain future. A woman counts her roubles outside a McDonald's restaurant which had opened in Moscow in January 1990, one of the earliest inroads made by Western consumer culture. McDonald's was the largest and best known global foodservice retailer with more than 24,500 restaurants in 16 countries by the end of the 1990s.

Taiwan, South Korea, Singapore, Malaysia and Hong Kong – followed the Japanese path, emerging as major manufacturers and exporters of consumer goods. They also proved adept at absorbing and adapting new technologies, particularly in high-growth sectors such as electronics. By 1995 Asia was producing scientists and engineers in numbers approaching those of the USA.

The post 1945 experience was different for other parts of the world. The centrally planned economies of the communist bloc followed the Soviet model of huge investment in heavy industry. Despite initial success and some impressive achievements in science, the system proved too rigid to absorb and generate new technologies, and increasingly fell behind the West. After the fall of communism in the late 1980s, Russia and the former Eastern bloc countries grappled with the painful task of readjusting to the demands of the market, with varying degrees of success. After the failure of Chairman Mao's "Great Leap Forward" – the industrialization drive around 1960 – China enjoyed more economic success in later decades by permitting a greater degree of private initiative.

In partly industrialized countries like Brazil and India, modern factories and primitive agriculture continued to exist side by side. Others, like the oil-producing members of OPEC, benefited when markets for their exports were favorable. A significant group of "Third World" countries, located in Africa, Asia and Latin America, continued to lag behind the rest of the world. With low standards of living, and poor access to education and healthcare, many in these "developing" nations in fact gained little from global developments. Some observers cited internal problems as the cause of continuing Third World poverty, but others blamed the richer nations for failing to tackle unequal international trading relations and the burden of debt.

Consumerism

Within the Western economies, the wealth generated during the 1950s was soon repairing the ravages of wartime. Improvements to the housing stock and education helped to raise the standard of living of the average citizen. In 1940 fully one-third of American homes lacked running water, but by 1970 this had fallen to only 1 in 87. Easily moulded and pigmented plastics, such as polyethylene and PVC, gave the world a new look, reducing production costs for a vast range of products. US consumption of plastics increased ten times between 1940 and 1960. The economies of mass-production insured that the prices of manufactured goods fell while new credit schemes enabled people to spend more. Initially, they bought

labor-saving consumer durables – electric irons, refrigerators, vacuu Continuous improvements in styling, design and performance steady stream of new products. Single-tub washing machines ga twin-tubs, which in turn were rendered obsolete by electronics-base ics. By 1990 nearly all British houses had a washing machi and blenders, fre microwave ovens washers added fu nological complex kitchen. But if o choose one machi

THE NEW 'ENGLISH ELE refrigera

Full of Fresh

Big science: information technology and biotechnology

The majority of the world's scientists who have ever lived live now. This was also true in 1900. In the single year 1995, sufficient new doctorates were awarded in science and engineering to populate a city the size of Camden, New Jersey, or Bath in the UK. The global expenditure on research and development was greater than the gross domestic product of Australia. It is thus hardly surprising that science was a major contributor to the technologies which transformed the industrialized world after 1945.

Communications and information technologies made striking progress and became major driving forces behind modern economies. High levels of government support had insured intensive research in computing from 1945. By the early 1960s, the valves of the early machines had been superseded by transistors. During the following decade, the use of semi-conductor technolo-

Bill Gates
founder of Microsoft, presenting prizes at the Science Museum in 1996; behind him an IBM PC and an Apple Lisa (left). The Internet offered a cultural and economic transformation of everyday life: a property auction held live over the Internet in 1999 (below). The first Internet auction had been held in 1995. Increasingly shopping, financial transactions and even the provision of music were shifting to the Internet. The first Internet bank, The Security First National Bank, was launched in October 1995.

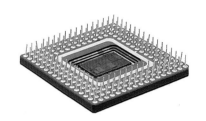

Information overload

An attempt to show, through the image of one man watching many screens, "zapper" in hand, the huge choice made available by the information revolution (right). Many predicted access to information to be one of the decisive divisions between the "haves" and "have-nots" of the future.

Genetics

A reconstruction o
model of the doub
of DNA, made by
James Watson in C
England, in 1953. T
represent the cher
"bases" whose ord
genetic code of an
The first animal wh
architecture" was d
mapped and seque
small nematode w
1998. Ten years ea
1989, this map (ab
the worm's chrom
produced, showing
groups of bases on

The personal computer

The Apple II, introduced in 1977 (above), was among the first. It sold for $1298. It succeeded the Apple I launched the year before which was the first cheap personal computer, though it was supplied on a circuitboard without a screen. Increasingly powerful chips put the power once reserved for large mainframe computers on the desks of individuals. The Intel 486 microprocessor introduced in 1989 incorporated a million transistors on a single silicon chip (above left). The founder of Intel, Gordon Moore, suggested in 1965 that the power of such chips would double every 18 months to two years. "Moore's Law" has continued to hold.

gies led to miniaturization, exemplified by the first mass-produced minicomputer, the DEC PDP8, launched in 1965. The late 1970s saw the introduction of personal computers, notably the Apple II (1977) and the Commodore PET. In 1981 IBM launched its first PC. Constant improvements to hardware and software allowed a wide range of new applications, and from the 1970s telephone and computer technologies began to converge. Simultaneous research into communications between computers, initially undertaken for military purposes, led to the formation of networks linking individual machines. Information-rich services had become the most prized commodity for both business and personal customers. At the end of the 20th century, car manufacturers – General Motors, DaimlerChrysler and Ford – were still the largest global corporations by revenue, but the meteoric rise of firms like software giant Microsoft and the microchip producer Intel reflected the increasing importance of information technology. It is perhaps significant that, while Herbert Hoover, US President in the 1920s, had been known as the "Great Engineer," his counterpart in the 1980s, Ronald Reagan, was called the "Great Communicator."

Healthcare improved thanks partly to the products of the pharmaceutical industry, and to biomedical engineering such as the imaging technologies of

Man's oldest problem: technology

advance of morality

JOHN FOWL

New Statesman, 6 August

opinion, particularly in Germany. The massive explosion in the Soviet nuclear reactor at Chernobyl in 1986 added to people's fears. Ironically it was a safety test at the plant that caused tons of radioactive material to be released into the atmosphere. In carrying out the test, operators had shut down vital safety equipment which would have prevented the accident. Emergency service personnel battling to contain the disaster died from exposure to radiation, and parts of Europe were affected by windborne radioactive dust from the plant.

live to t

at the e

Thro

for harr

life. Th

the subs

the cent

in futur

9 The age of ambiv

By the end of the 20th century, many countries had con famine and illiteracy as aberrations rather than inevitab ety. Yet as the pace of change accelerated, industrialized age of ambivalence – of mixed feelings about technology affected more areas of everyday life – from the food pec people worked and spent their leisure time – a growing about side-effects and unintended consequences undermin the entire concept of progress and those who promoted it.

Ambivalence towards technology was not new. Si Revolution, various groups and individuals had question of new technology. Whenever the pace of change seemed

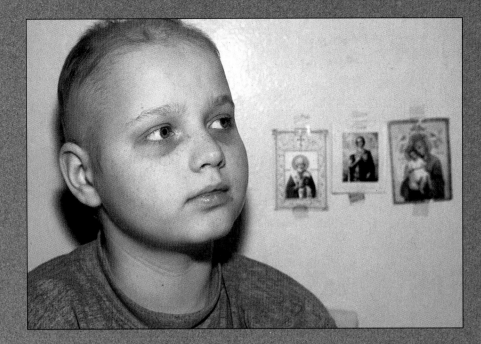

Chernobyl, Ukraine

at sunset. An almost idyllic view of the nuclear power station (below). Radioactive dust from the explosion of the Soviet designed RBMK-type reactor on April 26, 1986 polluted a large swathe of northern Europe. Helicopters dumped thousands of tons of sand, boron and lead, while untrained clean up teams went in without protective clothing and 116,000 people were evacuated. Among young people, the incidence of thyroid cancer grew ten times compared to the preceding five year period (right). The 200 tons of radioactive material in the reactor were subsequently encased in a sarcophagus of concrete. However, other reactors in the complex have continued in operation.

Environmental threats

First in Germany in the 1970s, and then elsewhere in Europe, an influential "green" lobby took shape, campaigning initially against nuclear power. By the end of the 1980s, words like "environmentalism" and the "green movement" became part of common parlance. Pressure groups like Greenpeace launched a series of campaigns to publicize various areas of concern. Overt politicization of the green debate went further in Europe, where official green parties scored a measure of success at local and national elections, than in the US.

Pressure groups argued that even routine levels of pollution posed a threat to the environment. In 1962, Rachel Carson's *Silent Spring* drew attention to the use of toxic chemicals in the countryside and the potential risks to wildlife. Concern about pollution caused by vehicle exhaust fumes in congested town centers was also widely expressed, and even the disposal of standard household waste posed ever greater logistical problems. Environmentally friendly technologies, ranging from the catalyst which cleaned car exhausts to home insulation, became the basis of major industries. High profile disasters attracted media attention. In 1984, thousands died as toxic fumes from the Union Carbide chemical works at Bhopal in India engulfed the town. Accidents involving oil tankers (including the *Torrey Canyon, Amoco Cadiz* and the *Exxon Valdez* in successive decades) caused havoc to the surrounding environment.

Combustion, internal and external

A communist icon, a Trabant car, is put to ironic use after German unification (left). Once the standard car produced in East Germany, its two-stroke engine was notoriously polluting. The Cypriot tanker *Haven* (right) on fire in the Gulf of Genoa in April 1991. It released over 140,000 tons of crude oil into the Mediterranean as it sank. In some such cases the chemicals used to clean the oil from affected beaches have accentuated rather than reduced the environmental impact of accidents. *Mad Dog 2* (inset) a solar-powered car developed by South Bank University in London. One of many projects seeking to develop pollution-free alternatives to gasoline burning cars, it reached 40km/h in the 1998 World Solar Rallye in Osaka.

Bhopal

Two of the quarter of a million people afflicted in the world's worst industrial accident. Toxic methylisocynate leaked from a tank at the Union Carbide pesticide plant on December 2, 1984. Some 50,000 people were hospitalized and 2,500 died.

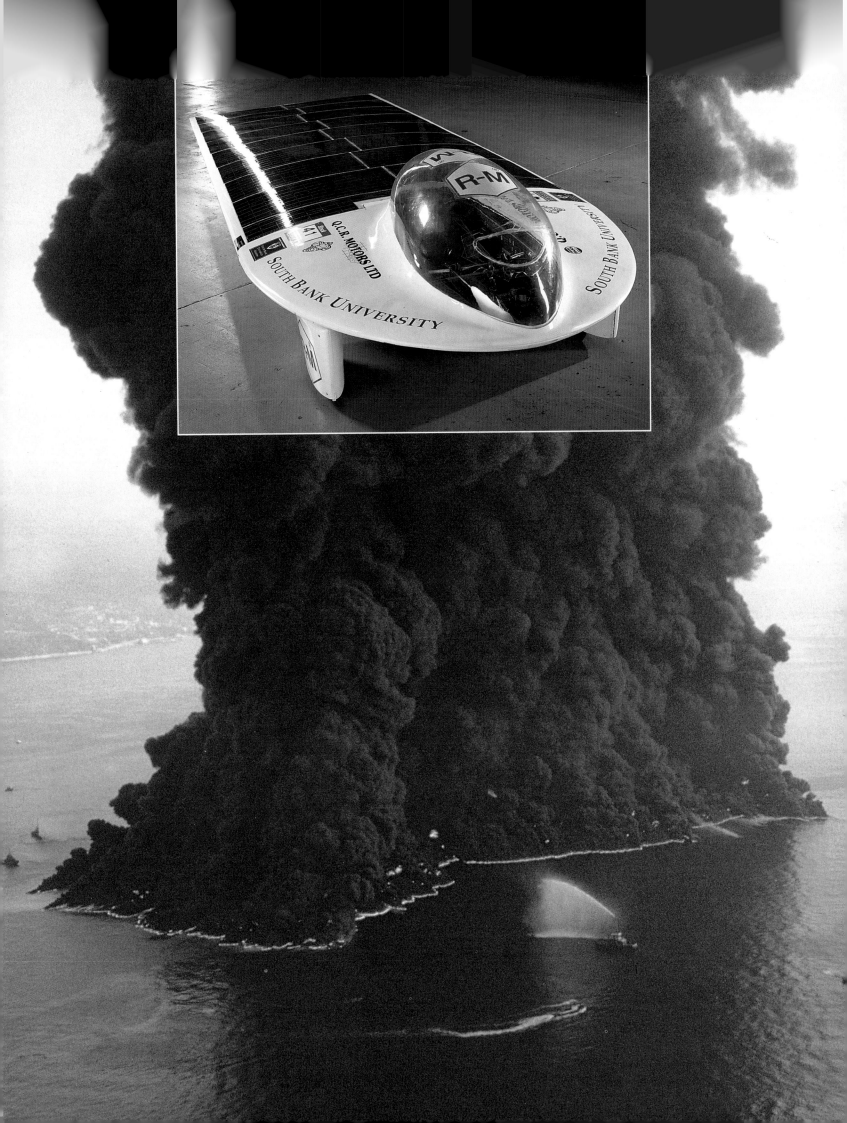

dioxide produced by burning fossil fuels; and the hole in the ozone layer, caused by the use of chlorofluorocarbons in aerosols and fridges. The detection of both depended on the most sophisticated analytical instrumentation. Furthermore, the solutions presented, whether alternatives to aerosol propellants or new energy sources, also drew on modern technology.

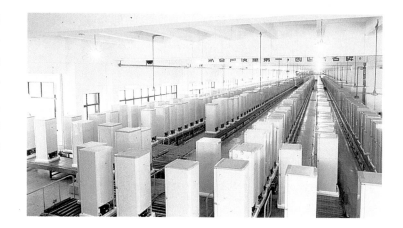

Ozone hole

The use of chlorofluorocarbons (CFCs) in refrigerators and aerosol sprays since the 1940s has been blamed for the thinning of the stratospheric ozone layer over the Antarctic, first reported in 1985. This "hole" has since widened and another has been observed over the Arctic. Ozone absorbs ultraviolet light which otherwise can cause skin cancer. It reacts with the products released by the breakdown of the CFCs in the stratosphere. In China's Kelon refrigerator plant (above left) in 1997 CFCs were still being used in some models for lack of an affordable alternative, although in the West such chemicals had been replaced (above).

Global warming

The Amazon rainforest is widely considered vital to the world's balance of carbon dioxide and so a key to global warming (left). In Brazil alone 2.5 million hectares were deforested in the years 1990–95 to provide farmland and timber, though the government did institute much stricter controls at the end of the 1990s. On the west coast of Antarctica, a 2 degree centigrade rise in temperature over the second half of the 20th century led to dramatic melting of the Larsen ice shelf (right), but exactly how this is related to global phenomena, manmade and natural, remains unclear.

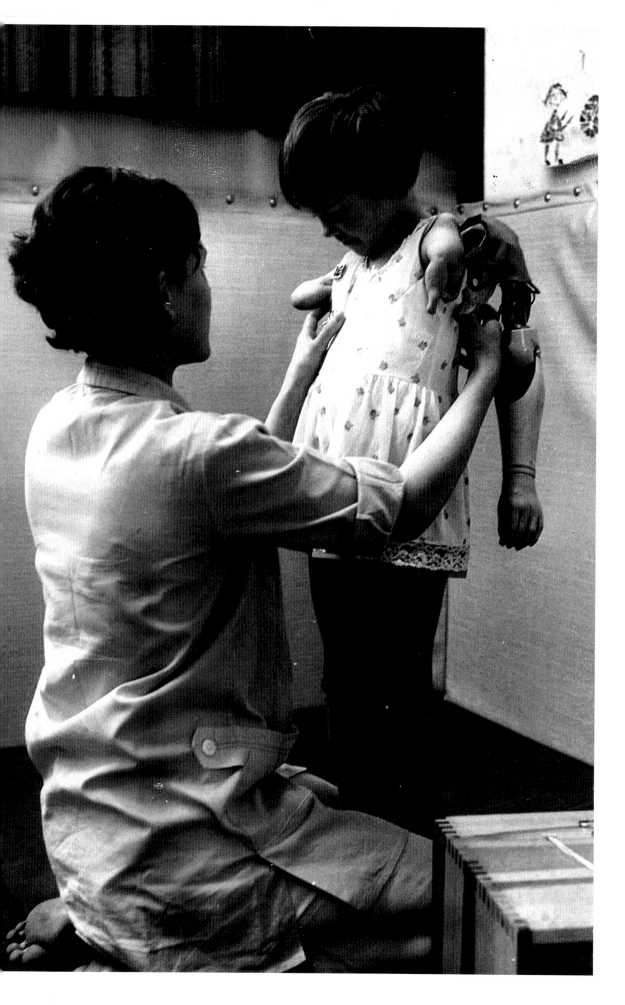

Medicines

A poster issued by the British National Health Service in 1999 (right) to dissuade patients from asking for antibiotics inappropriately. Many bacteria had become resistant to them partly because of misuse. Viagra tablets (far right). Released in the United States in March 1998, Viagra, sildanfil citreate, is prescribed to restore male sexual potency. Within seven months, six million prescriptions had been written.

Thalidomide

Fitting an artificial arm to a girl (left). About 10,000 children worldwide were born to mothers who used the sedative while pregnant. The drug was used in the years 1957 to 1962 in Germany (where it was developed), in the UK, Australia and several other countries, though it was never licensed for general use in the USA. The tragedy precipitated a radical tightening in the procedures for approving drugs, particularly in the USA, a national campaign for compensation in the UK and a prosecution of company executives in Germany. Ironically, the drug was later found to be a valuable treatment for leprosy and certain symptoms of AIDS.

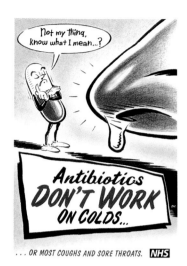

Medicine was another area where the application of science and the latest technology came under a cloud, in the eyes of some at least. Faith in the pharmaceutical industry was partly undermined by the thalidomide disaster in the late 1950s and early 1960s, when a sedative offered to pregnant women caused deformities in unborn children. Many people became noticeably wary of the claims of the drug companies. In Britain in the late 1990s, one adverse report about possible side-effects of the contraceptive pill led to a brief panic during which the number of unwanted pregnancies rose markedly. In the same period, however, when Viagra seemed to offer men enhanced sexual performance, global demand was huge even before the drug had been licensed in many countries. While many remained sceptical as to the safety or effectiveness of proprietary medicines, "alternative" medicine, ranging from homoeopathic and herbal remedies to quasi-mystical practices, flourished as never before.

Anxieties about new technology could often reflect distrust of politicians and corporations: hostility towards genetically modified foods was fuelled by suspicion of the commercial motives behind their promotion, while genetic research conjured up fears of "designer babies" and other practices darkly reminiscent of eugenics.

Meditation

A session in progress at the Findhorn Foundation, one of the most famous alternative, "new-age" communities in northeast Scotland (below).

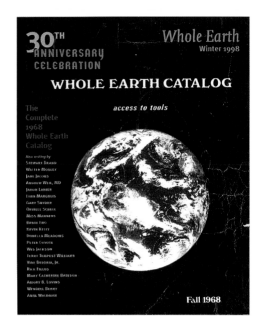

If governments and firms attracted much criticism, individuals could also be highly selective in the ways they responded to the potential risks of technology. As early as 1968, the *Whole Earth Catalog* offered the growing community of environmentally-aware consumers a range of everyday products. Increasingly, "organic" foods produced without recourse to chemicals, "free-range" eggs and meat from animals raised in humane conditions, and a large variety of vegetarian and vegan foods appeared on sale. Initially expensive and confined to specialist outlets, they eventually appeared in mainstream supermarkets, although the higher price still discouraged many potential purchasers. An ever growing range of "ethical" products became available,

Whole Earth Catalog
a reprint of the original, first published in 1968 (left). It was the unofficial counterculture handbook for those in search of environmentally sound goods.

Organic produce
sold in Austria with all the professionalism devoted to mainstream foods (below).

GM foods

Some oilseed rape (right, May 1999) and soya beans, (top right) have been genetically modified. Genes have been implanted from other organisms to confer resistance to specific pests or diseases. Opponents of GM foods, such as the Greenpeace movement, have proposed that there are potential dangers in introducing alien characteristics into living organisms: Greenpeace activists destroying GM crops in France (above).

UNITED COLORS
OF BENETTON.

Full circle
Commercial firms have been able
to turn public concern about
environmental dangers to their own
advantage, using it in advertisements,
first to shock people into paying
attention, then to develop their
image and so to sell more products.
A winter 1992/93 Benetton
advertisement (left) using an image
of an oil-soaked seabird, actually a
casualty of Iraq's destruction of
Kuwaiti oil wells, not of a tanker
spillage.

from toys produced by local co-operatives to sophisticated investment port-
folios. Consumer action was occasionally able to influence the behavior of
firms and retailers, but was rarely applied to great effect. People
might have responded to greener technologies offering clear
savings in money terms: drivers switched to smaller cars in periods
of high fuel prices, and bought unleaded gasoline where this was
subject to lower taxes. But where the financial benefits were not so
obvious, few were willing to put themselves out with a whole-
hearted commitment to recycling, energy saving, or products with
an ethical slant.

AGAINST
ANIMAL
TESTING

another
HAPPY EASTER bunny

THE BODY SHOP

185g GUARANTEE
This product has
selected and imp
and should reach
condition. If it d

RINCES

NA CHUNKS

BRINE

DOLPHIN FRIENDLY

Radical chic

The fashion store fcuk was adopting a similar approach to Benetton, by using protective suits to advertise their wares, in April 1997 (right).

Green credentials

Drawing on public sentiment The Body Shop advertised (left) that it would not use ingredients tested on animals for cosmetic purposes after December 31, 1990. Dolphin friendly tuna, caught without the use of drift nets which often accidentally catch and kill dolphins (far left). During the 1980s approximately 50,000 km of driftnets were out nightly. In 1989 the United Nations agreed to ban their large-scale use on the high seas.

Late 20th century technologies

Aberfeldy Footbridge in Scotland, 1992: the world's first cable-stayed footbridge made entirely of advanced composite materials, which offer lightness and durability, and require little maintenance.

The Intel Pentium microprocessor, 1993: a huge leap in computing power, with over 3 million transistors.

Easi-Breath inhaler, from Norton Healthcare Ltd, 1999. A measured dose of asthma medication is released when the patient inhales.

The City-Cabrio from MCC smart, with a short body, convertible roof and low fuel consumption, designed for use in Europe's crowded cities.

Pentax SFX camera: an all-automatic single lens reflex camera launched on the British market in 1987.

Boeing 747, first introduced in 1970, the most successful commercial airplane ever, and still the largest civil airliner.

The Fiat Pendelino, the Italian high-speed tilting train introduced in 1988: an artist's impression of the train in Virgin livery.

Hubble space telescope: a 2.40 metre reflecting telescope orbiting 600 km above the Earth since 1990, capable of producing images with a resolution ten times higher than that of ground-based telescopes.

The Dyson Dual Cyclone DC06 vacuum cleaner, 1999: with over 50 sensory devices and three computers, it can clean rooms on its own, avoiding obstacles and stairs.

Lara Croft, the digitally generated heroine of the Tomb Raider series of computer games.

Scanning tunneling microscope, 1986, by WM Technology of Cambridge: for analyzing the surfaces of solid objects, with sufficient accuracy to identify individual atoms.

Sony Memory Stick Walkman, announced in 1999: a digital audio playback device equipped with a memory card, allowing the distribution of music via the Internet.

Backdrop: Earthrise, a classic view of the Earth above the Lunar horizon taken by the astronauts on board Apollo 10 in 1969.

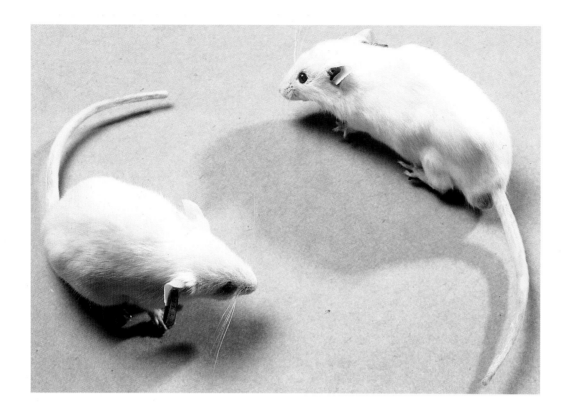

Transgenic animals

Descendants of the first mammals to carry a US patent. Each mouse carries a transgene which comes partly from a virus. It has a unique genetic sequence which is not found in mice in the natural world. The mouse, also known as an "oncomouse," is genetically modified to develop cancer, making it a useful tool in research into the disease. On the other hand, several groups oppose the patenting of a mouse designed for experimental purposes.

Epilogue

As the world started the next Millennium, it was clear that future prosperity depended on the judicious application of existing and new technologies, together with closer scrutiny and controls over those which might pose risks to the environment or human life. The integration of broadcasting, telecommunications and computing technologies was bringing unbounded opportunities for electronic commerce and new forms of communication and entertainment. Biotechnology was offering new ways forward for healthcare and agriculture, but also challenges to traditional assumptions of what is "natural," and about personal identity. The excitement and anxiety at the end of the 20th century offered curious parallels to the anticipations at the beginning. The option of simply rejecting technology as bad, in favor of some idealized pretechnological world, had not been convincing even when mooted in the late 19th century. Now, thousands of millions of people aspired to the standard of living they could see on global television networks. At the same time, there was at least a rhetorical consensus that governments, firms and consumers shared a responsibility to ensure that technologies would not be used in ways that could endanger the future of the human species. This was so general that it was hard to realize its novelty.

As our 18th century forebears first concluded, science and technology raise questions that cannot be confined to or answered only in the factory and the laboratory. They pose challenges and opportunities for society as a whole. Their history is an intrinsic part of all our history.

World Wide Web

The NeXT computer (below) on which Tim Berners-Lee worked at the European nuclear physics center, CERN, as he designed the World Wide Web with Robert Cailliau from 1989. An image of Tim Berners-Lee (right) composed of 2,304 web pages, which first appeared in *Time* magazine in March 1999. This was the fourth in a series on the most influential people of the century.

Further reading

(Where a book relates to particular chapters, these are indicated in brackets after the reference.)

Bauer, Martin (ed.), *Resistance to New Technology: nuclear power, information technology and biotechnology*, Cambridge University Press, 1995 (chapter 9)

Biggs, Lindy, *The Rational Factory: architecture, technology and work in America's age of mass production*, Johns Hopkins University Press, 1996 (chapters 5 and 6)

Blume, Stuart, *Insight and Industry: on the dynamics of technological change in medicine*, MIT Press, 1992 (chapter 8)

Bud, Robert and Gummett, Philip (eds), *Cold War, Hot Science: applied research in Britain's defence laboratories, 1945–1990*, Harwood Academic Publishers, 1999 (chapter 7)

Cardwell, D. S. L., *The Fontana History of Technology*, Fontana, 1994

Chandler, Alfred D., *The Visible Hand: the managerial revolution in American business*, Harvard University Press, 1977 (chapters 5 and 6)

Chant, Colin (ed.), *Science, Technology and Everyday Life, 1870–1950*, Routledge, 1989 (chapters 5 and 6)

Cooter, Roger, *The Cultural Meaning of Popular Science: phrenology and the organisation of consent in nineteenth-century Britain*, Cambridge University Press, 1984 (chapters 1 and 3)

Cowan, Ruth Schwartz, *More Work for Mother: the ironies of household technology from the open hearth to the microwave*, Free Association Books, 1989 (chapters 6 and 8)

Derry, T. K. and Williams, T. I., *A Short History of Technology from the Earliest Times to AD 1900*, Clarendon Press, 1960

Durant, J., Bauer, M. W. and Gaskell, G., (eds), *Biotechnology in the Public Sphere: a European sourcebook*, Science Museum, 1998 (chapter 9)

Edwards, Paul N., *The Closed World: computers and the politics of discourse in Cold War America*, MIT Press, 1996 (chapter 7)

Freeman, Michael, *Railways and the Victorian Imagination*, Yale University Press, 1999 (chapter 4)

Hamlin, Christopher, *Public Health and Social Justice in the Age of Chadwick, 1800–1854*, Cambridge University Press, 1998 (chapter 3)

Hecht, Gabrielle, *The Radiance of France: nuclear power and national identity after World War II*, MIT Press, 1998 (chapter 7)

Heppenheimer, T. A., *Countdown: a history of space flight*, Wiley, 1997 (chapter 7)

Hounshell, David A., *From the American System to Mass Production, 1800–1932: the development of manufacturing technology in United States*, Johns Hopkins University Press, 1985 (chapters 2 and 6)

Hughes, Thomas P., *American Genesis: a century of invention and technological enthusiasm, 1870–1970*, Penguin, 1990

Hughes, Thomas P., *Networks of Power: electrification in Western society, 1880–1930*, Johns Hopkins University Press, 1983 (chapters 5 and 6)

Inkster, Ian, *Scientific Culture and Urbanisation in Industrialising Britain*, Ashgate, 1997

Jennings, M. L. and Madge, C. (eds), *Pandaemonium 1660–1886: the coming of the machine as seen by contemporary observers*, Deutsch, 1985

Kenwood, A. G. and Lougheed, A. L., *The Growth of the International Economy, 1820–1990: an introductory text*, 3rd edn, Routledge, 1992 (chapters 5, 6 and 8)

LeMahieu, D. L., *A Culture for Democracy: mass communication and the cultivated mind in Britain between the wars*, Clarendon Press, 1988 (chapter 6)

Luckin, Bill, *Questions of Power: electricity and environment in inter-war Britain*, Manchester University Press, 1990 (chapter 6)

Marx, Leo, *The Machine in the Garden: technology and the pastoral ideal in America*, Oxford University Press, 1964 (chapter 3)

Mathias, Peter and Davis, John A. (eds), *The First Industrial Revolution*, Blackwell, 1992 (chapters 1 and 2)

Mayne, Alan, *The Imagined Slum: newspaper representation in three cities, 1870–1914*, Leicester University Press, 1993 (chapter 3)

Mokyr, Joel, *The Lever of Riches: technological creativity and economic progress*, Oxford University Press, 1992 (chapters 1, 2, and 5)

O'Brien, Patrick K. and Quinault, Roland (eds), *The Industrial Revolution and British Society*, Cambridge University Press, 1993 (chapters 1 and 2)

Pagnamenta, Peter and Overy, Richard, *All our Working Lives*, British Broadcasting Corporation, 1984 (chapters 6, 7 and 8)

Palfreman, Jon and Swade, Doron, *The Dream Machine: exploring the computer age*, British Broadcasting Corporation, 1991 (chapters 7, 8 and 9)

Pick, Daniel, *War Machine: the rationalisation of slaughter in the modern age*, Yale University Press, 1993 (chapters 6 and 7)

Pursell, Carroll, *White Heat: people and technology*, British Broadcasting Corporation, 1994

Smith, Denis (ed.), *Perceptions of Great Engineers: fact and fantasy*, Science Museum, 1994 (chapter 4)

Travis, Anthony S., *The Rainbow Makers: the origins of the synthetic dyestuffs industry in Western Europe*, Associated University Presses, 1993 (chapter 5)

Weart, Spencer R., *Nuclear Fear: a history of images*, Harvard University Press, 1988 (chapter 9)

Picture credits

Abbott Laboratories, 198TL
Advertising Archives, 190B
American Philosophical Society, 20L
Arcaid/Richard Bryant, 186L, 187L
BASF Corporate Archives, Ludwigshafen am Rhein, 102T, 102B
Benetton Group SpA/Photographed by Steve McCurry. Concept: Oliviero Toscani, 216T
The Boots Company Archive, 154T
Bridgeman Art Library/© DACS 2000, 127T
British Film Institute/© CTE (Carlton) Ltd, 156L
Bulletin of the Atomic Scientists, 202BM
Camera Press, 168TL
Canadian Centre for Architecture, Montreal, 90TL
John Chillingworth, 175TR
Company Archive, Harrods Ltd, London, 120M
Core Design Ltd, 219MR
Deutsches Museum, Munich, 96-7T
Dyson Ltd, 219ML
Environmental Images, 204–5 both (Clive Shirley), 207 main (Paolo Vaccari), 210B (Irene Lengui), 211L (Vanessa Miles) 211R (Steve Morgan), 215TR (Martin Bond), 215B (Toby Adamson)
Mary Evans Picture Library, 63T
Ford Motor Company, 140B, 141T
French Connection Limited, 217
Greenpeace, 210T (Hindle), 214B (Einberger/Argum), 215TL (Rouvillois)
The Guardian, 191B (Sean Smith), 196B (Martin Argles), 209T (Graham Turner), 213B (Denis Thorpe)
Ernst Haas Studios/Hulton Getty Picture Collection, 182T
Hagley Museum and Library, 45B
Andrea Hitzler, 206T
Hulton Getty Picture Collection, 14L, 25B, 30B, 35B, 38, 39, 48–9 both, 50b, 51 all, 52, 53T, 53M, 55M, 56T, 57T, 60 both, 61 both, 62 both, 63B, 64BL, 64BR, 65B, 66 both, 67M, 67B, 68 both, 70B, 71 all, 76B, 77TR, 77B, 78ML, 81TL inset, 83T, 84–5T, 84M, 86, 87 both, 90ML, 90B, 91M, 91B, 93 both, 94–5 both, 101TL, 101TM, 101TR, 101ML, 101M, 101MR, 101BL, 101BR, 103MR, 103BL, 105L, 107T, 108TL, 109BR, 111TL, 111B, 112B, 114B, 115 all, 116B, 117B, 119, 120B, 121TL, 122B, 124–5 main, 126M, 126B, 127B, 128 both, 129 both, 132, 133 both, 134 both, 136–7 main, 137TR inset, 138 both, 139TR, 142 both, 143, 146T, 149, 150TR, 150BR, 151TR, 151BL, 153, 154B, 156R, 157B, 160B, 161, 162BL, 162BR, 163TL, 163TR, 166T, 167, 168BL, 168BR, 169TL, 169TM, 169ML, 169M, 169MR, 169BL, 169BM, 169BR, 170B, 171 all, 173 all, 174 all, 175ML, 176T, 177B,
177TR, 178–9 main, 180T, 188BL, 188BR, 188R, 189TL, 190T, 198B, 200, 202T, 202BL, 203, 208, 212
Hulton Getty Picture Collection/Crown copyright is reproduced with the permission of the Controller of Her Majesty's Stationery Office, 177TL
Imperial War Museum, 166B
Wilhelm Lewicki, 27T
Liaison Agency/Bartholomew, 206B
Manchester Daily Express/Science & Society Picture Library, 172
Manchester Public Library, LSU, 13
MCC smart, 218MR
AC Michael, 126T
MSP, Maunsell Ltd, 218TL
NARA, National Archives at College Park, 162T
NASA/Hulton Getty Picture Collection, 162T
NASA/Science & Society Picture Library, 218–9 main, 219TR
© National Gallery, London, 42 main, 80–1 main
The National Gallery of Scotland, 75
National Gas Archive, 64TL
National Maritime Museum, Greenwich, London, 15R
National Museum of Photography, Film & Television/Science and Society Picture Library, 5, 65T, 113BL, 113BR, 140T, 144 both, 145BL, 145TR, 146T, 146B, 152T, 155, 185, 191TM, 218BL
National Portrait Gallery, London, 25T
National Railway Museum/Science & Society Picture Library, 78T, 78MR, 79 both, 81TR inset, 82 all, 118, 159
© The Natural History Museum, 24
The Print and Picture Collection, The Free Library of Philadelphia, 121M, 121B
Public Record Office, 54T, 55T
Records of the Patent and Trademark Office, National Archives, RG 241, 44
Retrograph Archive Limited, 183T
Sainsbury's Archive, 121TR
Science Museum/Science & Society Picture Library, 1, 2, 3, 6, 8, 9, 10–11, 12 both, 14R, 15I, 16 all, 17, 18 both, 19, 20B, 21, 22–3 all, 26, 27B, 28, 29, 30T, 31 both, 32T, 32BL, 32BM, 34T, 34BL, 35T, 35M, 36 both, 37 both, 40, 41 all, 42L inset, 42R inset, 43, 45T, 45M, 46, 47, 50TL, 50TR, 54B, 55B, 56BL, 56BR, 57M, 57BL, 58 both, 59 both, 64TR, 67T, 69L, 70T, 72, 73, 74 all, 76T, 77TL, 78B, 80t inset, 83B, 88 all, 89 all, 91T, 92, 96B, 98, 99, 100 both, 101BM, 102M, 103TL, 103TR, 103ML, 103BR, 104 all, 105TR, 105BR, 106 both, 107B, 108ML, 108MR, 108B, 109TL, 109TR, 109BL, 110 both, 112T, 124T, 125T inset, 130, 131, 135, 137TL inset, 139TL, 139BL, 139BR, 145M, 145BR, 152B, 157T, 158, 163B, 164–5 all, 169TR, 176M inset, 180B, 181 both, 184, 188T, 189R, 190MR, 191TL, 191TR, 193B, 195TR, 196TL, 197MR, 197L, 198TR, 198BR, 199 both, 201, 207T inset, 213TR, 216BL, 218TR, 218ML, 219BL, 220 both
Science Museum/Science & Society Picture Library/© DACS 2000, 150I
Science Museum/Science & Society Picture Library/Crown copyright is reproduced with the permission of the
Controller of Her Majesty's Stationery Office, 176B
Sears, Roebuck and Co, 120T
Robert Silvers/www.photomosaic.com, 221
Slava Katamidze/Hulton Getty Picture Collection, 147 all, 168TR
Sony UK Ltd, 219BR
Stone London, 186R (Doug Armand), 187R (Steven Weinberg), 194T (Gary John Norman), 194B (Wayne Eastep), 197R (John Edler), 209B (Jon Riley), 218BR (Mark Wagner)
Topham Picturepoint, 170T
Trustees of the Wedgwood Museum, Barlaston, Staffordshire (England), 20T
Universal/Ronald Grant, 151TL
University of East Anglia, 175B
US Library of Congress/Hulton Getty Picture Collection, 32BR, 84B, 85B, 111TR, 112M, 113T, 114T, 116T, 117T, 117M, 122T, 123, 141B, 146M, 148 both, 160T, 193T
V&A Picture Library, 33, 97B
Jason Vass Gallery, Santa Monica/The Coca-Cola Company 195TL
Virgin Trains, 219TL
Wellcome Institute Library, London, 53B, 57BR, 69R
Whole Earth Review, 214T
© Wilse/Norwegian Folk Museum, 107M

We regret that we have been unable to trace the copyright owners of the images on pages 126T, 192, 195TR, 195B, 206T, 213TL and 216MR to request permission for their use.

Index